Manual of United States Surveying.

SYSTEM

OF

RECTANGULAR SURVEYING

EMPLOYED IN SUBDIVIDING THE

PUBLIC LANDS OF THE UNITED STATES;

ALSO

INSTRUCTIONS FOR SUBDIVIDING SECTIONS AND RESTORING LOST CORNERS OF THE PUBLIC LANDS.

Illustrated with Forms, Diagrams and Maps;

CONSTITUTING A

COMPLETE TEXT-BOOK OF GOVERNMENT SURVEYING.

FOR THE USE OF U. S. DEPUTY SURVEYORS, COUNTY SURVEYORS, AND ALL WHO CONTEMPLATE ENTERING THE PUBLIC SURVEYING SERVICE.

TO WHICH IS ADDED

AN APPENDIX

CONTAINING INFORMATION IN REGARD TO ENTERING, LOCATING, PURCHASING AND SETTLING LANDS UNDER THE VARIOUS LAND LAWS, ETC. ETC.

BY J. H. HAWES,

LATE PRINCIPAL CLERK OF SURVEYS IN THE GENERAL LAND OFFICE.

PHILADELPHIA:

J. B. LIPPINCOTT & CO.

1873.

PREFATORY REMARKS.

THE following pages have been prepared to supply a want which is widely known and felt. The General Land Office is constantly receiving letters from county surveyors and others, soliciting information in regard to the system of government surveying, how to subdivide sections, restore missing corners, etc. etc.

It is the custom of the department, in answer to these varied inquiries, to furnish as explicit directions as can be given within the ordinary limits of an official communication; but it is quite impracticable in such a communication to set forth in detail the principles and the laws, with their multifarious bearings and applications, which affect or control the surveyor in restoring obliterated public surveys, or running and marking the boundaries of legal subdivisions not before established in the field; and yet this information is essential to the surveyor who would execute his work correctly and in accordance with law.

In view of the great number of inquiries of this character received by the General Land Office, the writer, who for several years had especial charge of the department of government land surveying, commenced the preparation of a circular to be printed for the use of the office, which should be sufficiently comprehensive to meet the class of inquiries referred to.

The examination and reflection incident to the preparation of such a circular continued to develop new complications, and suggest new points to be explained, enlarging the scope of the work and ultimately inducing a change in the original design.

It became apparent that the only plan which would afford surveyors all the information necessary to enable them to discharge their duties properly, was not only to lay down specific rules in particular cases, but to give a full and complete exposition of the surveying system, and the laws and instructions relating thereto. To give directions how to restore a lost corner without also affording some insight into the laws and practice under which it was originally established, would be like giving a theoretical explanation of a difficult field operation in civil engineering to one not conversant with that branch of mathematical science

The government system of surveying is in some respects peculiar and unlike any other, and no adequate facilities have been afforded surveyors not employed in the public service to make themselves acquainted with its rules and principles. Hence it is in many cases impracticable to make instructions intelligible to the local surveyor, without first giving some explanations as to the manner in which the public surveys are executed.

The writer has been frequently and forcibly impressed with this truth, when endeavoring to relieve correspondents of embarrassments occasioned by a want of the proper knowledge in regard to the laws and practice of the government surveying system. Where these are understood the instructions become comparatively simple and are readily comprehended.

In 1855, a manual of instructions to regulate the field operations of United States deputy surveyors, was prepared and printed under the direction of the General Land Office. Other instructions had been printed at earlier periods, but the manual prepared in 1855 embraced all the improvements suggested by the experience of the surveying department up to that time, and was much more comprehensive and complete than anything of the kind which had preceded it.

By the second section of an act of Congress, approved May 30th, 1862, it is provided: "That the printed manual of instructions relating to the public surveys, prepared at the General Land Office, and bearing date February 22d, 1855, the instructions of the Commissioner of the General Land Office, and the special instructions of the Surveyor-General when not in conflict with said printed manual or the instructions of said Commissioner, shall be taken and deemed to be a part of every contract for surveying the public lands of the United States."

A supplemental pamphlet containing many recent changes

authorized by the department was prepared by the writer for the use of the office, and printed July 1st, 1864. The modifications and additions contained in this pamphlet are now made a part of every surveying contract, subject of course to such subsequent changes as may from time to time be found necessary.

These rules and instructions are only intended to be furnished by the department to persons in the government service, and the aim of the writer has been to embody them in a practical form for the use of the student and the general public; he has endeavored to furnish a TEXT-BOOK which will enable surveyors to fit themselves for the public service. To this end the present work has been systematized and arranged in convenient divisions and subdivisions under appropriate heads, and illustrated by examples, forms, diagrams and maps.

To the Manual proper are added instructions for restoring extinct lines and corners of the public surveys, and for subdividing sections. These directions are based upon the laws of Congress and the well-established precedents of the General Land Office, and are very complete, embracing, it is believed, the principles at least, of nearly every case that will arise in practice. County surveyors in the public land states know well the difficulties experienced in executing this kind of surveys, and will find these instructions a great aid in performing their work.

The manner of proceeding to have certain public lands surveyed in advance of the regular progress of the public surveys, under the provisions of the 10th section of the act of May 30th, 1862, known as the deposit system, is explained; also how to proceed to get certain small islands surveyed.

An Appendix is added to the work containing instructions, forms, and rulings of the General Land Office, in regard to entering, locating, purchasing, and settling lands under the several acts following, to wit: Pre-emption Laws, Homestead Laws, Military Bounty Act, Agricultural College and Revolutionary Land Scrip, Mining Laws, etc.

Also in regard to the adjustment of public grants of land to states and corporations, for railroads, canals, schools, universities and other purposes, and the survey and settlement of private land claims, under foreign titles and special acts of Congress, with various forms, etc.

The comprehensive index attached to the book will enable the reader to turn readily to any subject desired.

The writer cheerfully acknowledges his indebtedness to HON. JAMES M. EDMUNDS, late Commissioner of the General Land Office, for his valuable aid in selecting and preparing the instructions contained in the Appendix. He also takes this occasion to return his acknowledgments to HON. JOSEPH S. WILSON, the present Commissioner, for the uniform courtesy and kindness shown him in affording every facility to obtain the necessary data to complete the work and bring it down to current date.

Trusting that he has to some extent succeeded in producing a hand-book of the public land system suited to the wants of a large and intelligent class of his fellow-citizens, the writer respectfully submits the following pages to the public, with the earnest wish that they may be instrumental in securing a uniformity in the laws and practice in the several land states, conforming to the laws of Congress and the decisions of the United States Courts.

TABLE OF CONTENTS.

APPENDIX.

THE PUBLIC LAND SYSTEM.

THE public land system of the United States was inaugurated as early as the year 1785, and by the experience of many years has been brought to great perfection. It is conducted upon the same comprehensive and liberal principles which distinguish all the beneficent institutions of the government—institutions looking always to the benefit of the many, favoring especially the middle classes and the poor, and that stand forth as imperishable monuments of the profound wisdom and enlarged philanthropy of the great statesmen who organized and established them.

The scope and magnitude of the land system of the United States will be more fully comprehended by a brief retrospect of its history.

The territory northwest of the Ohio and east of the Mississippi rivers, known as the Northwestern Territory, being within the limits of the United States as defined by the treaty of 1783, which terminated the revolutionary war, was conveyed to the general Government, with certain reservations, by the State of New York, in 1781; by Virginia, in 1784; by Massachusetts, in 1785; and by Connecticut, in 1786. The territory acquired by these several cessions embraced the extent of country which constitutes the States of Ohio, Indiana, Illinois, Michigan, Wisconsin, and that part of Minnesota east of the Mississippi river.

In 1802, Georgia conveyed to the United States that

portion of the present States of Mississippi and Alabama lying north of the 31st degree of north latitude.

By the treaty of 1803, the United States acquired from the French Republic the Territory of Orleans and Louisiana, embracing that portion of the States of Mississippi and Alabama south of the 31st degree of north latitude, and the vast country which now constitutes the States of Louisiana, Arkansas, Missouri, Iowa, Kansas, Oregon, and that part of Minnesota west of the Mississippi river; also the Territories of Nebraska, Dakota, Montana, Idaho, Washington, and the Indian Territory.

The State of Florida was acquired by treaty with Spain in 1819, and the States of California and Nevada, and the Territories of New Mexico, Arizona, Colorado, and Utah were obtained by treaty with the Republic of Mexico in 1848.

Extending over 23 degrees of latitude and stretching away over 45 degrees of longitude, this vast public domain embraces eighteen states and nine territories, the smallest of which contains an area greater than that of England and Wales, and the largest twice as many square miles as the whole of Great Britain.

The aggregate area of this vast extent of country is 1,446,716,072 acres, of which 485,311,778 acres had been surveyed by the government prior to the 1st day of July, 1867. In Ohio, Indiana, Illinois, Michigan, Iowa, Missouri, Mississippi, Arkansas, and Alabama, the public surveys have been finished excepting some small islands and fragmentary strips of land along the margins of rivers, bayous, etc. The surveys are also nearly completed in Wisconsin, Louisiana, and Florida.

Within the territory acquired by the United States there were numerous private claims, the titles to which were derived from the former governments. Where, upon proper investigation, the titles have been found to be valid, these claims have been segregated from the public lands and patented to the claimants, the area of such private grants already confirmed and patented amounting in the aggregate to 14,770,351.

The public domain was first opened to public sale and then to private entry. Congress, in 1830, passed a law allowing actual settlers to secure titles to their locations on very liberal terms and without being subjected to compete at auction with outside bidders. The preemption system underwent various modifications from 1830 down to 1841, growing more and more liberal and beneficent towards the settler, until 1862, when the whole public domain, surveyed and unsurveyed, offered and unoffered, was thrown open to actual settlement at the lowest price and upon the most accommodating terms. Under the operation of this law the enterprising pioneer could select his homestead of 160 acres from the choicest fields of this extended country and commence his improvements without having to pay anything for his land until one year after the public surveys were extended over it, and with a certainty that his rights would be recognized and respected by the government.

But there was yet another step to be taken in the direction of a still more liberal policy towards the western pioneer. Congress, in 1862, passed the Homestead Law, giving to each settler who entered upon and cultivated the land, 160 acres, for which he has only to pay ten dollars to the government and four dollars as compensation to the officers for making out the necessary papers.

The acts of the National Legislature have uniformly encouraged emigration and afforded the industrious poor man an opportunity to secure for himself a farm and a comfortable home. Upwards of four hundred millions of acres of as choice land as can be found in any country have passed to the ownership of individuals, and some fifteen hundred thousand families are to-day living in the enjoyment of comfortable, and thousands of them of luxurious homes, acquired from the government under the operation of this benign policy.

But the liberality in public gratuities has been even more munificent than that towards individuals. More than 7,000,000 acres have been given to the states to aid in educating the people, thereby giving strength and vigor to the elements of a free government.

Then Congress conceded to certain states the "swamp lands" within their limits, the aggregate selections thus far reaching 60,246,532 of acres. Upwards of sixty-five millions of acres have been granted for military services in the revolutionary war and the war with Great Britain, Mexico, etc. Five hundred thousand acres have been granted to each of the land states for internal improvements, grants to universities, and to aid in the building of railroads.

Finally, Congress, in 1863, granted 150,000,000 of acres to aid in building the Pacific Railroad, a scheme which is destined at no very distant day to consolidate the industrial and commercial interests of the country, open up and people the vast empire beyond the western limits of the present frequented paths of civilization, and bind together the Atlantic and Pacific States, with their intermediate links, in a bond of union which internal conspiracies or foreign cupidity and avarice shall never be able to break, but which the lapse of ages shall weld stronger and stronger while time lasts.

The policy of Congress with reference to the public domain, as indicated by their acts, appears to be not to look to the public lands as a source of direct revenue, but rather by encouraging *bona fide* settlements and aiding important works of internal improvement and institutions for the education of the people, to quicken the settlement of the country and the development of its resources, and, by increasing individual wealth and education, secure national riches, prosperity and happiness, thereby realizing the true end and aim of a democratic government, based upon the virtue, intelligence and freedom of its people.

THE PUBLIC LAND DEPARTMENT.

The Constitution of the United States delegates to Congress the power of disposing of the public domain. The General Land Office is the executive bureau, under the supervision of the Secretary of the Interior, to carry into effect the laws of Congress relating to the public lands, and direct the various details of the public land system. The chief officer of the General Land Office is denominated the Commissioner, to whom all communications intended for this bureau should be addressed.

SURVEYING DISTRICTS.

The states and territories in which there are unsurveyed public lands are divided by Congress into surveying districts. Some of these districts comprise one or more states or territories, but no more than one district is made out of a single state or territory. A Surveyor-General is appointed by the President of the United States, by and with the advice and consent of the Senate, for each surveying district, whose duties are to direct the surveying operations in his district, under the instructions of the General Land Office.

The following table shows the surveying districts as at present constituted, and the location of each Surveyor-General's office, together with the name and salary of each of said officers.

TABLE I.

Surveying Districts.	Surveyor-General's Office—where located.	Name of Surveyor-General.	Annual Salary.
California and Arizona....	San Francisco.......	Lauren Upson....	$3000
Nevada....	Virginia City........	A. P. K. Safford.	3000
Oregon........................	Eugene City..........	E. L. Applegate..	2500
Washington Territory......	Olympia..............	S. Garfield.........	2500
Colorado and Utah	Denver, Col..........	Wm. H. Lessig...	3000
New Mexico..................	Santa Fé, N. M......	John A. Clark....	3000
Kansas.........................	Leavenworth, Kan.	Hiram S. Sleeper	2000
Iowa and Nebraska.........	Plattsmouth, Neb...	P. W. Hitchcock.	2000
Minnesota.....................	St. Paul..............	Levi Nutting.......	2000
Dakota........................	Yankton..............	Wm. Tripp........	2000
Idaho...........................	Boise City............	L. F. Carteé......	2500
Montana........................	Helena................	S. Meredith.......	2500
Louisiana	Not in operation....	....	2000
Florida.........................	Not in operation....	....	2000

Note.—The public surveys have been completed in the following states, excepting some small islands and fragmentary strips of land along the margins of rivers, bayous, etc., to wit: Ohio, Indiana, Illinois, Michigan, Iowa, Wisconsin, Missouri, Mississippi, Arkansas, and Alabama. Nearly all the public land is also surveyed in Louisiana and Florida.

Each Surveyor-General is allowed a chief clerk, draughtsman, and such number of clerks as are necessary to carry on the business of his office, all of whom receive their appointments from said Surveyor-General. If the fact were generally understood that the letting of all contracts for surveying the public lands and all positions in said offices are under the immediate and sole control of the Surveyors-General, fewer communications soliciting employment in the public surveying service would be addressed to the General Land Office, and some perplexity and disappointment would be saved to such applicants.

RATES PAID FOR PUBLIC SURVEYING.

All surveying contracts are made at specified rates per lineal mile of line "actually run and marked in the field," offsets and random lines not included, the rate varying according to the kind of line to be run and the locality in which the survey is to be made.

The following are the surveying rates authorized by law for the respective states and territories:

TABLE II.

Names of States and Territories.	Section Lines.	Standard and Township Lines.	Meridian and Base Lines.
California and Arizona........	$12 00	$12 00	$15 00
Nevada..............................	12 00	12 00	15 00
Kansas..............................	5 00	6 00	10 00
Iowa and Nebraska.............	5 00	6 00	10 00
Oregon...............................	7 00	8 00	8 00
Washington Territory..........	......	......	......
Colorado............................	7 00	8 00	10 00
Utah..................................	12 00	12 00	20 00
Minnesota..........................	......	......	......
Dakota Territory.................	5 00	6 00	10 00
New Mexico.......................	7 00	8 00	10 00
Arizona..............................	10 00	12 00	15 00
Idaho.................................	......	......	......
Montana.............................	......	......	......
Louisiana and Florida.........	4 00	4 00	4 00

The prices in the foregoing table are the maximum rates allowed by law, but the full prices are not always paid. In some localities surveying can be done at less than these rates, and still afford a fair compensation to the surveyor for his labor. In the more remote districts, where the cost of provisions and labor is greater, the full rates are allowed, and in some instances higher rates than those named are paid where the appropriation is based upon estimates at greater prices.

When the field work is completed, the field notes are returned to the office of the Surveyor-General, where they are examined and tested, and plats* and transcripts are prepared and transmitted to the General Land Office,

* The plats are uniformly constructed on a scale of 40 chains to an inch.

approved by the Surveyor-General in his official capacity, together with the surveying account of the deputy; and no surveys are paid for until such plats and transcripts are examined at the General Land Office and found correct. The plats and transcripts are prepared without cost to the deputy surveyors.

TABLE III.

Areas of the States and Territories.

Public Land: States and Territories.	Areas of the Public Land States and Territories.	Public Lands surveyed to June 30, 1865.	States.	Area. Acres.
Ohio	25,576,960	25,576,960	Maine............	22,400,000
Indiana........	21,637,760	21,637,760	New Hampshire	5,939,200
Michigan......	36,128,640	36,128,640	Vermont.........	6,535,680
Illinois.........	35,462,400	35,462,400	Massachusetts..	4,992,000
Wisconsin.....	34,511,360	34,511,360	Rhode Island....	835,840
Iowa............	35,630,898	35,228,800	Connecticut......	3,040,000
Minnesota.....	53,459,840	22,045,867	New York........	30,080,000
Missouri.......	41,824,000	41,824,000	New Jersey......	5,324,800
Arkansas......	33,406,720	33,406,720	Pennsylvania....	29,440,000
Mississippi....	30,179,840	30,179,840	Delaware.........	1,356,800
Louisiana.....	26,461,440	23,461,440	Maryland........	7,119,360
Alabama.......	32,462,080	32,462,080	Virginia..........	39,265,280
Florida.........	37,931,520	26,631,520	North Carolina..	32,450,560
Kansas.........	52,043,520	16,171,776	South Carolina..	21,760,000
Nevada.........	71,737,741	728,119	Georgia...........	37,120,000
California......	101,717,362	27,680,685	Kentucky.........	24,115,200
Oregon.........	60,958,720	5,730,186	Tennessee........	29,184,000
Nebraska......	48,636,800	13,561,132	Texas.............	175,587,840
Dakota.........	153,982,080	1,859,989	Dis't Columbia..	38,400
New Mexico...	77,568,640	2,293,142		
Arizona........	72,906,304			476,584,960
Colorado.......	66,972,292	1,622,251		1,446,716,072
Utah............	56,355,635	2,425,239		
Montana.......	92,016,640		Total......	1,923,301,032
Idaho...........	58,196,480			
Washington...	44,796,160	3,530,645		
Indian Ter'y..	44,154,240			
*	1,446,716,072	474,160,551		

* Russian America, acquired by purchase in 1867, is estimated to contain 369,529,600 acres.

The following is a comparative statement of the land measures of the United States and the French measures formerly used in the late Province of Louisiana. These proportions were adopted by the Surveyor-General's Office of St. Louis, Missouri, and are considered in all surveys as the true proportions between said measures.

TABLE IV.

Comparative Land Measures.

LINEAR MEASURE.

French feet.		U. S. feet.
72	are equal to	77
Perches.		Poles.
6		7
	Chains.	Links.
1	0	29 166
2	0	58 333
3	0	87 5
4	1	16 661
5	1	45 833
6	1	75
7	2	04 166
8	2	33 333
9	2	62 5
10 or 1 arpent lineal,	2	91 666
2	5	83 333
3	8	75
4	11	66 666
5	14	58 333
6	17	50
7	20	41 666
8	23	83 333
9	26	25
10	29	16 666
100	291	66 666
1000	2916	66 666
12	35	
Side of a league square.		
84	245	
Side of a mile square.		
Arpents.	Perches.	
27	4 2-7	80

SUPERFICIAL MEASURE.

Arpents.		Acres.
288	are equal to	245
1		0 85 07
2		1 70 14
3		2 55 21
4		3 40 28
5		4 25 35
6		5 10 42
7		5 95 49
8		6 80 56
9		7 65 625
10		8 50 69
100		85 06 94
1000		850 69 44
10000		8506 94 44
Arpents.	Perches.	Acres.
1	17 551 =	1
2	35 102 =	2
2	52 653 =	3
4	70 204 =	4
5	87 755 =	5
7	05 306 =	6
8	22 857 =	7
9	40 804 =	8
10	57 959 =	9
11	75 510 =	10
117	55 102 =	100
1175	51 020 =	1000
11755	10 204 =	10000
A league square contains		
Arpents.		Acres.
7056		6002 50
A mile square contains		
Arpents.	Perches.	
725	32 64	640

A league square contains 7056 arpents or 6002 50 acres. A mile square contains 725 arpents and 32·64 perches, or 640 acres.

12 arpents = 35 chains lineal.

TABLE V.

Location of United States Land Offices.

OHIO.

Chillicothe.

INDIANA.

Indianapolis.

ILLINOIS.

Springfield.

MISSOURI.

Booneville,
Ironton,
Springfield.

ALABAMA.

Mobile,
Huntsville,
Montgomery.

MISSISSIPPI.

Jackson.

LOUISIANA.

New Orleans,
Monroe,
Nachitoches.

MICHIGAN.

Detroit,
East Saginaw,
Ionia,
Marquette,
Traverse City.

ARKANSAS.

Little Rock,
Washington,
Clarksville.

FLORIDA.

Tallahassee.

IOWA.

Fort Des Moines,
Council Bluffs,
Fort Dodge,
Sioux City.

WISCONSIN.

Menasha,
Falls of St. Croix,
Stevens' Point,
La Crosse,
Bayfield,
Eau Claire.

CALIFORNIA.

San Francisco,
Marysville,
Humboldt,
Stockton,
Visalia,
Sacramento.

NEVADA.

Carson City,
Austin,
Belmont.

WASHINGTON TER.

Olympia,
Vancouver.

MINNESOTA.

Taylor's Falls,
St. Cloud,
Winnebago City,
St. Peter,
Greenleaf,
Du Luth.

OREGON.

Oregon City,
Roseburg,
Le Grand.

KANSAS.

Topeka,
Junction City,
Humboldt.

NEBRASKA.

Omaha City,
Brownsville,
Nebraska City,
Dakota City.

NEW MEXICO TER.

Santa Fé.

DAKOTA TER.

Vermillion.

COLORADO TER.

Denver City,
Fair Play.

IDAHO TER.

Boise City,
Lewiston.

MONTANA TER.

Helena.

ARIZONA TER.

Prescott.

SYSTEM

OF

RECTANGULAR SURVEYING

EMPLOYED IN SUBDIVIDING

THE PUBLIC LANDS OF THE UNITED STATES.

INTRODUCTION.

The rectangular system of surveying adopted by the United States in subdividing the public lands, in its present state of perfection, is the simplest and most beautiful that could be devised. It is believed no other government equals our own in the perfection of its system of public surveys.

A state when subdivided has the regularity and symmetry of a well-laid out city on a grand scale; the townships corresponding to the blocks and the sections and subdivisions to the lots, but with this difference in favor of the public surveys—the sections and townships are uniformly one and six miles square, bounded by lines conforming to the cardinal points.

Then there is the principal meridian and base line, crossing each other at right angles, which form the frame-work upon which the subsequent surveys are built, answering to the main streets of a city, from which the blocks are consecutively numbered; so that any one possessing a knowledge of the system, can determine the locality and relative position of a township or subdivision with as great facility and certainty as he can a block or lot in a well-planned town.

Imagine one vast city extending over 50,000 acres, surveyed in this manner, under the general supervision of one directing head, and we may have some just conception of the regularity and beauty of the government system of rectangular surveying.

This perfection was not reached at once; the existing system is the result of many years' trial and change. The first government surveying was done in the State of Ohio, in 1796. The "seven ranges," as they were called, being the first seven tiers of townships west of the Ohio river, were first surveyed. The land department was at that time under the direction of the

Secretary of the Treasury,* Hon. Albert Galatin, and the work was executed under the immediate supervision of Rufus Putnam, the first Surveyor-General. The townships in these seven ranges commence respectively with number one on the river and number north. The present mode of reckoning townships and ranges was not adopted until a recent date.

The convergency of the meridians was soon found to be a source of serious difficulty where long range lines were projected. Several plans were tried to overcome this difficulty, resulting finally in adopting the "correction parallels." Parallels are now required to be run every five townships or thirty miles north or south of the base line.

Improvement has been gradual and slow; modifications have gone on step by step, until the present symmetrical and perfect system has been attained. We may now, therefore, safely venture to put forth a Hand-Book on this subject, feeling confident that the facilities thereby for the first time afforded the general public for becoming acquainted with the system of government surveying, will promote the public interest, and be appreciated by all who are interested in the subjects treated of.

---

* The General Land Office was organized as a distinct bureau of the Treasury Department by act of Congress, approved April 25th, 1812.

SYSTEM

OF

U. S. RECTANGULAR SURVEYING.

---

1. TOWNSHIPS.—The public lands of the United States are primarily surveyed into uniform rectangular tracts, six miles square,* called *Townships*, bounded by lines conforming to the cardinal points, and containing, as nearly as may be, 23,040 acres.

2. SECTIONS.—The townships are subdivided into thirty-six tracts, one mile square, called *Sections*, containing (except in cases hereinafter explained) 640 acres each.

The sections are numbered consecutively from one to thirty-six, beginning at the northeast corner of the township and numbering west with the north tier of sections, thence east with the second tier, west with the third tier, and so on to section thirty-six in the southeast angle of the township. (Fig. 1.)

FIG. 1

6	5	4	3	2	1
7	8	9	10	11	12
18	17	16	15	14	13
19	20	21	22	23	24
30	29	28	27	26	25
31	32	33	34	35	36

3. SUBDIVISIONS OF SECTIONS.—Sections are divisible into four equal parts of 160 acres each, called *Quarter Sections*,

---

* See Standard Parallels.

and each quarter section is again divisible into two *half-quarter sections* of 80 acres, or four quarter-quarters containing 40 acres each. (Fig. 2.) These are called *Legal Subdivisions*, and are the only divisions recognized by the government in disposing of the public lands, except where tracts are made fractional by water-courses or other causes, or in the case of town lots.*

Fig. 2.

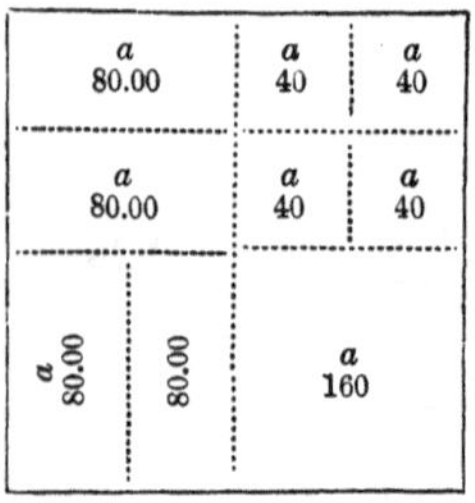

[The subdivisions of sections are not actually surveyed and marked in the field. Quarter section or half mile posts are established on the boundaries of the sections, and the quarter-quarter corners are by law the equidistant points between the section and quarter section corners; but the interior subdivisional lines of sections are made only on the plats of townships, at the Surveyor-General's office; and when the boundaries of these subdivisions are required to be established on the ground, a county surveyor or other competent person is employed.]

4. Principal Meridians and Base Lines.—Two principal lines are established prior to the survey of the townships—a north and south line denominated a *Principal Meridian*, and an east and west line styled a *Base Line*. These lines constitute the basis of the public surveys, and are prerequisite to the laying out of townships.

5. Ranges.—Any number or series of townships situated in a tier north and south are denominated a *Range*, and the ranges are designated by numbers east or west,

* In some of the old land states public surveys have been made which did not conform to the rectangular system. Lots were surveyed with given frontages on rivers, bayous, etc., and running back to such depth as would embrace the required areas, regardless of the cardinal points. Such surveys were made by authority of special enactments, and were exceptions to the established system of rectangular surveying.

as the case may be, from the governing meridians. The townships in each range are also numbered north or south from established base lines.

6. STANDARD PARALLELS.—Townships are said to be six miles square, but the law requiring that the north and south lines shall conform to the true meridian, it is evident that in consequence of the convergency of the meridians, these lines will continue to approach each other as they are extended northward, thereby throwing the townships out of square. To correct this convergency, and preserve as nearly as practicable the square form of the township, *Standard Parallels*—called also and more appropriately *Correction Lines*—are run every five townships or 30 miles north and south of the base line.*

These parallels or correction lines are run due east or west, and constitute new bases for the townships north of them, up to the next parallel or base line.

COURSING, MEASURING, AND MARKING LINES.

1. BURT'S SOLAR COMPASS.—Deputy surveyors are required to use Burt's improved solar compass or other instrument of equal utility, in surveying standard and township lines; but when the needle can be relied on, the ordinary magnetic compass may be used in subdividing or meandering.

2. STANDARD CHAIN.—The chain used in the field must be carefully compared from day to day with a *Standard Chain* furnished by the Surveyor-General, to be carried along by the deputy; and any variation in the length of the chain in use, from the opening of the links or other cause, must be promptly corrected.

3. TALLY PINS.—The deputy surveyor will use eleven

* Standard parallels were formerly run every 24 miles north of the base line, and every 30 miles south of it. The present system was adopted in 1866.

tally pins, made of steel, not exceeding fourteen inches in length, weighty enough toward the point to make them drop perpendicularly, and having a ring at the top, in which is fastened a piece of red cloth, or something else of conspicuous color.

4. Marking Tools.—The best marking tools adapted to the purpose should be procured, and all letters and figures should be distinctly and neatly cut. A rat-tail file and a small whetstone will be found indispensable articles to keep the marking tools in order.

5. Horizontal Measurement.—The length of all lines must be ascertained by horizontal measurement, taking care always to keep the chain stretched to its utmost tension. In ascending or descending steep hills or mountains, the chain may have to be shortened to half its length or even shorter in order to obtain the true horizontal measure. Care must also be taken to have the tally pins properly plumbed.

6. Process of Chaining.—In measuring lines with a two-pole chain, every five chains are called "a tally," because the last of the ten tally pins with which the forward chainman set out will have been stuck. He then cries "tally," which cry is repeated by the other chainman, and each registers the tally by slipping a button, ring of leather, or something of the kind, on a belt worn for the purpose. The hind chainman then comes forward, and having counted in the presence of his fellow the tally pins which he has taken up, so that both may be assured that none have been lost, he takes the forward end of the chain and proceeds to set the pins. The chainmen continue to change places alternately, so that one is forward in all the odd and the other in all the even tallies. It is believed this plan will most surely prevent a mis-tally.

7. Line Trees.—Trees immediately in line are marked by two chops or notches on each side, and are called "*line trees*," "*station trees*," or "*sight trees*."

8. Marking Lines.—A sufficient number of other trees nearest the line on either side must be *blazed* on two sides quartering toward the line, the blazes to coincide with the direction of the line, and to approach each other the further the line passes from the blazed trees. The line should be so conspicuously marked as to be readily followed.

Where trees two or more inches in diameter are found, the required blazes must not be omitted.

Bushes on or near the line should be bent *at right angles therewith*, and receive a blow of the ax at about the usual height of blazes from the ground, sufficient to leave them in a bent position, but not to prevent their growth.

9. Insuperable Objects on Line.—When insuperable objects are met with on line, such as ponds, lakes, marshes, rivers, etc., they must be passed by taking the necessary right-angle offset, or by traverse or trigonometrical operation, all the particulars of which must be stated in the field book. "Meander posts" must be set at the intersection of the line with the pond or other obstacle, on both margins, and the course and distance therefrom given to two trees in opposite directions. These trees must be marked with a blaze and notch facing the post. On the margin of navigable lakes or water-courses the fractional section, township and range must also be marked upon the trees.

10. Random Lines.—Trees should *not be blazed* in running *random* or *trial lines*. Bushes and limbs may be lopped, and stakes may be set at every ten chains, to enable the surveyor to follow the line on his return, but the stakes must be pulled up when the true line is established.

When bushes or limbs are lopped, they should be bent *in the direction of the line*, to prevent mistaking random for true lines.

11. Lines how Run.—All north and south lines, except meridians and lines between the north tier of sections in

the township, are run *from* south *to* north; true east and west lines, except base lines, correction parallels, and lines between the west tier of sections in the township, are run *from* east *to* west.

PERPETUATING CORNERS.

1. The *chief purpose of the public surveys* is to establish the corners of the public lands; and however true the coursings and accurate the measurements, the principal object will not be attained if the corners are not made permanent. The importance of perpetuating all corners of the public surveys in the most durable manner cannot be overrated.

2. The principal corners established in government surveys are of four kinds, to wit: 1, *Township Corners;* 2, *Section Corners;* 3, *Quarter-Section Corners;* and 4, *Meander Corners;* and four different modes are employed to perpetuate them, respectively, depending upon certain conditions, as follows:

(1) *Corner Trees.*—When a tree not less than five inches in diameter stands immediately in place, it is found to be the best means of perpetuating *any description of corner* that can be employed, and should be adopted in preference to all others.

(2) *Stone Corners.*—Where suitable stones can be readily procured, the deputy surveyor is required, in all cases except when a tree is found, to prefer them before either of the other modes of perpetuating corners, as constituting the most durable monument it is practicable to erect.

Stones used for corners must have a length of at least 14 inches, and contain not less than 500 cubic inches.

All corner stones 14 inches long or more, and less than 18 inches in length, should be set two-thirds of their length in the ground; if more than 18 inches long, they should be set three-quarters of their length in the ground.

(3) *Posts and Witnesses.*—It frequently happens in a timbered country that suitable stones cannot be obtained. When this is the case, and trees are at hand for "witnesses," posts may be planted at the corners, and evidenced as directed under the head of "bearing trees" on page 32. All posts must be made of the most durable wood of the forest at hand.

In loose or alluvial soil, section, quarter-section, or meander posts may be *driven* into the ground, instead of digging holes and planting them; but no posts should be so driven unless, from the nature of the soil, they will be thereby rendered more firm and enduring.

(4) *Posts and Mounds.*—Where neither stones nor witness trees are to be found, the corners must be marked by mounds of earth erected around posts. This is the common mode of perpetuating the corners on prairie lands.

3. Manner of Constructing Mounds.—The mode of erecting posts and mounds is as follows: The post will

Fig. 3.

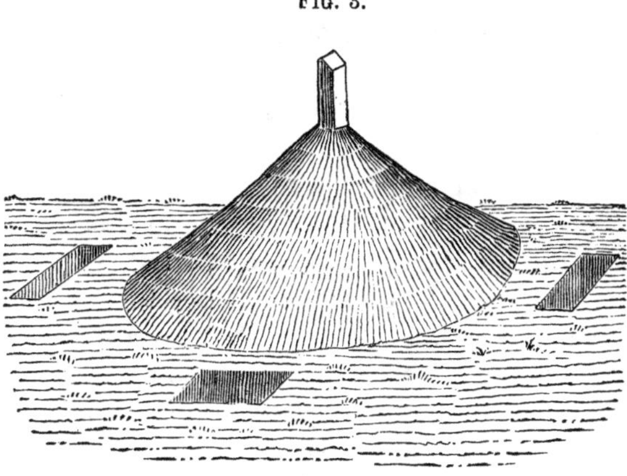

be planted or driven into the ground to the depth of 12 inches at the precise corner point; and a marked stone, a small quantity of charcoal or a charred stake must be deposited 12 inches below the surface—against the north side of the post, when the deputy is running north; against the west side, when he is running west, etc.— as witnesses in the future. It is optional with the surveyor to adopt either of these memorials; but one of them must in every case be employed, and the deputy must state in the field book *which* is used.

Having planted the post—which must be of the same dimensions as prescribed for the same kind of corner *without* the mound—dig four "*pits*" at least 12 inches deep, on opposite sides and 6 feet from the post, piling and closely packing the excavated earth around it in such a manner as to form a cone-shaped mound, leaving the post to project 12 inches above the apex. Where sod is to be had, the mound must be covered with it, grass side up; but sod must never be wrought up with the earth in forming the mound. (Fig. 3.)

The pits should be located at right angles, and at a uniform distance from the center point; but where it is found necessary, owing to the impracticable nature of the soil, to dig the pits or either of them at a greater distance than 6 feet, and in other directions, the course and distance to each pit so located must be noted in the field-book.

4. Bearing Trees.—The position of all corner posts or corner trees, of whatever description, must be witnessed by taking the courses and distances of two or more adjacent trees in opposite directions, as hereinafter directed. These bearing trees are distinguished by a smooth blaze facing the corner, with a notch at its lower end; and in the blaze is inscribed the number of the township, range and section. The letters "B. T."—bearing tree—are also cut upon a small blaze directly under the other, and as near the ground as practicable. If the tree should be a beech, or other smooth, firm bark, the

marks may be made on the bark and the blazes may be omitted.

Where a tree not less than 2½ inches in diameter can be found for a bearing tree within 300 links of the corner, it should be used.

5. Witness Pits.—Whenever the requisite number of witness trees cannot be found, the deficiency must be supplied by digging pits 2 feet square and not less than 12 inches deep.

STANDARD AND CLOSING CORNERS.

Two sets of corners are established on standard parallels and base lines; one when said lines are run, and the other when the exterior and subdivision lines on the south of them are closed thereon. These corners are separately explained below; instructions for perpetuating and marking them will be found in their appropriate place.

1. Standard Corners.—At the time the parallels and base lines are run, the township, section, and quarter section corners are established thereon. As the township and section lines north are run *from* them, it follows that these corners will be common to two townships, sections, or quarter sections north of the parallel or base line, and these are called *Standard Corners.*

2. Closing Corners.—North and south lines are required to be run on the true meridian. Hence, when the township and section lines below reach the parallels or base lines north, they will not close on the standard corners previously established, because of the convergency of the meridians, but will strike the line at a distance corresponding to the convergency; east of the standard corners if the field of operations be west of the governing meridian, and west of said corners if the surveys be east of the principal meridian. Another set of township and section corners is therefore established at

the points of intersection with said standard or base line, and the distances of said corners from the corresponding standard corners previously set, are measured and noted in the field book. The corners so established are called *Closing Corners*, and will of course be common to two townships or sections *south* of said base or standard line. No closing quarter section corners are established. (Fig. 4.) (See Quar. Sec. Cor., p. 38.)

FIG. 4.—Standard and Closing Corners.

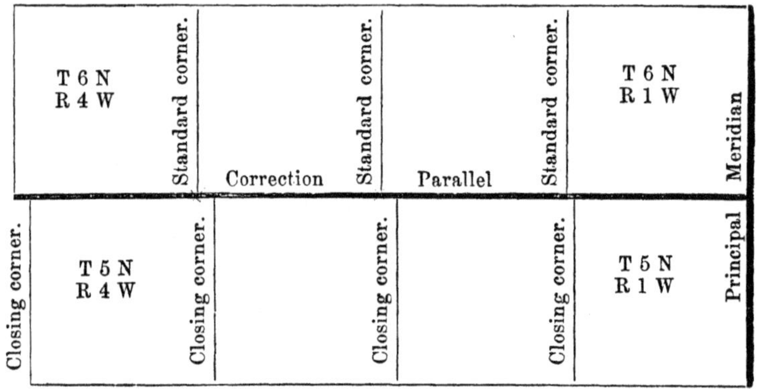

T 6 N
R 4 W
Standard corner.
Correction
Standard corner.
Parallel
Standard corner.
T 6 N
R 1 W
Principal Meridian
Closing corner.
T 5 N
R 4 W
Closing corner.
Closing corner.
Closing corner.
T 5 N
R 1 W

TOWNSHIP CORNERS.

Township corners are established at intervals of 6 miles each, and are perpetuated by the following modes, to wit:

1. TOWNSHIP CORNER POSTS.—The *post* is placed first in order because circumstances render its use most common in practice. Corner posts are required to be 4 feet in length, and at least 5 inches in diameter, and are to be planted to the depth of 2 feet, the part projecting above the ground being squared to receive the marks required to be cut upon them.

When the corner is common to four townships, the post is set cornerwise to the lines, presenting the angles to the cardinal points, and on each flattened side must be marked the number of the township, range and section which it faces. Thus, if the post be common to townships one and two south of the base line, and one and two west of the meridian, it should be set and marked as indicated by Fig. 5. Six notches will also be cut on each of the four edges. (Fig. 6.)

Fig. 5.

N.

T. 1 S. R. 2 W. S. 36.	T. 1 S. R. 1 W. S. 31.
T. 2 S. R. 2 W. S. 1.	T. 2 S. R. 1 W. S. 6.

W. — E.

S.

(a) *Standard Township Corners.*—If the post is on a standard parallel or base line, and is common to only *two* townships on the north side thereof, six notches will be cut in the east, north, and west edges, and the letters "S. C." (Standard corner) will be cut on the flattened surface, but no notches will be cut in the south edge. (See Standard Corners, p. 33.)

Fig. 6.

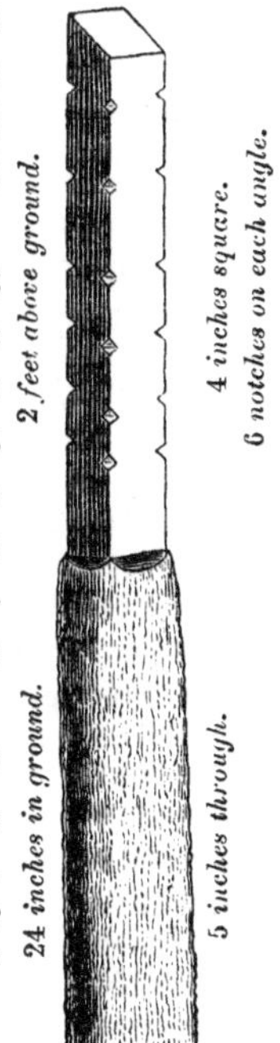

(b) *Closing Township Corners.*—If the post is common to two townships south of the parallel or base line, six notches will be cut in the east, south, and west edges, but none in the north edge, and the letters "C. C." (Closing corner) must be cut upon the flattened surface. The manner of establishing standard and closing corners is explained on page 33.

The position of *all* township corner posts must be witnessed by four bearing trees, one in each of the adjoining townships, marked in the manner prescribed under the head of "Bearing trees," or by "pits," where trees cannot be found.

2. Township Corner Stones.—Township corner stones must be inserted in the

ground not less than eight inches, with their sides to the cardinal points, and small mounds of stones should be constructed against the sides of them.

The notches on the edges are the only marks required, and the directions for notching township *posts* are to be followed in notching corner stones in like circumstances. (Fig. 7.)

Fig. 7.

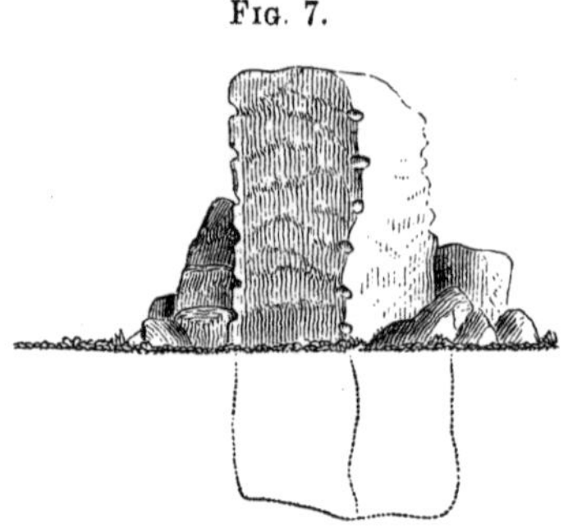

3. Township Corner Trees.—A tree in place, when employed to perpetuate a township corner, must be marked and witnessed in the same manner as a township *post.*

4. Township Corner Mounds.—The post and mound is a common method of marking corners. The manner of constructing the mound is explained on page 31. Mounds at township corners must be 5 feet in diameter at their base, and 2½ feet in perpendicular height. Posts in township mounds, therefore, require to be 4½ feet in length, so as to be planted 12 inches in the ground, and allow 12 inches to project above the mound.

The pits for a township mound will be 18 inches wide, 2 feet in length, and at least 12 inches deep, located 6 feet from the post, and on opposite sides.

At corners common to *four* townships, the pits will be placed on the lines and lengthwise to them. On base and parallel lines, where the corners are common to only *two* townships, *three* pits only will be dug—two in line on either side of the post, and one on the line north or south of the corner, as the case may be. By this means the standard and closing corners can be readily distinguished from each other.

Posts in mounds should be notched, marked, and faced precisely as directed for posts without the mound.

SECTION CORNERS.

Section corners are established at intervals of 1 mile or 80 chains, and the four modes of perpetuating corners already described, are employed to mark them, to wit:

1. Section Corner Posts.—Posts for section corners must be 4 feet in length and 4 inches in diameter, firmly planted or driven into the ground to the depth of 2 feet, the part projecting being squared to receive the required marks. (Fig. 8.)

Fig. 8.

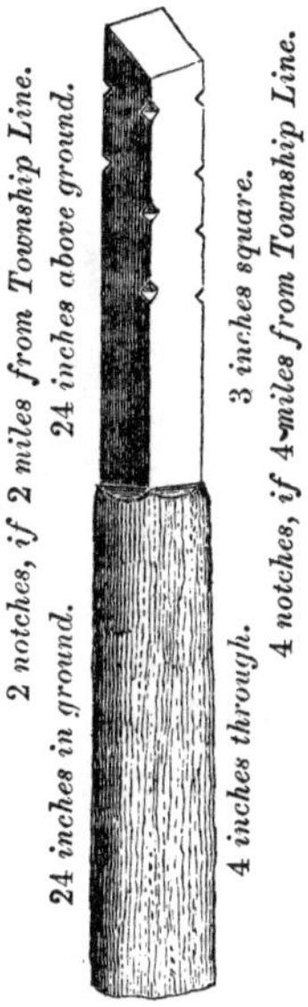

When the corner is common to four sections, the post will be set *cornerwise* to the lines, and on each flattened surface will be marked the number of the section which it faces; also, on the northeast face, the number of the township and range will be cut.

All mile posts *on township lines* will have as many notches on the two corresponding edges as they are miles distant from the respective township corners. Section posts *in the interior of a township* will have as many notches on the *south* and *east* edges as they are miles from the south and east boundaries of the township, but *no* notches on the north and west edges. By this plan the corner can be identified thereafter, if the post be found lying upon the ground.

All section posts, whether in the interior of a township or on a township line, must be witnessed by four bearing trees, one in each of the adjoining sections, to be marked in the manner described under the head of "Bearing trees."

When the requisite number of bearing trees cannot be found, the deficiency will be supplied by substituting pits 18 inches *square*, and not less than 12 inches in depth.

2. SECTION CORNER MOUND.—Mounds at section corners will be $4\frac{1}{2}$ feet in diameter at their base, and 2 feet in perpendicular height; the post being 4 feet in length and inserted 12 inches in the ground. The post must be not less than 3 inches square, and is to be marked and witnessed the same as the post without the mound.

At corners common to four sections, the post in mound will be set with the edges to the cardinal points; at corners common to only two sections, the flattened sides of the post will face the cardinal points.

FIG. 9.

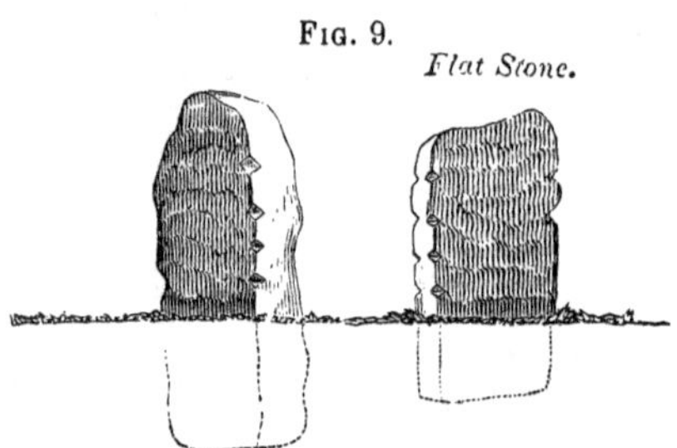

3. SECTION CORNER STONES.—When stones are used for section corners on township lines, they will be set with their edges in the direction of the line; but when standing for interior section corners they will be planted facing the north, and should be notched the same as section posts similarly situated. No marks except the notches are required, but they will be witnessed by trees or pits as required where posts are used. The requisite dimensions of corner stones are given under the appropriate head. (Fig. 9.)

4. SECTION CORNER TREE.—A tree in place at a section corner is marked according to the direction for marking section posts.

QUARTER SECTION CORNERS.

Quarter section corners are established at intervals of half a mile or 40 chains, except in the north and west tiers of sections in a township. Where the section lines exceed or fall short of 80 chains, in subdividing these sections, the quarter post is established just 40 chains from the interior section corner, throwing the excess or deficiency upon the last half mile. The intervals between the quarter posts and the north and west township boundaries will therefore be irregular.

Quarter section corners are not required to be established on the *north* boundary of the northern tier of sections in a township *south of and bordering on a standard parallel or base line.* The resurvey of standard or base lines, by the deputy surveyor, for the purpose of establishing such quarter posts, is unnecessary and will not be paid for.

Fig. 10.

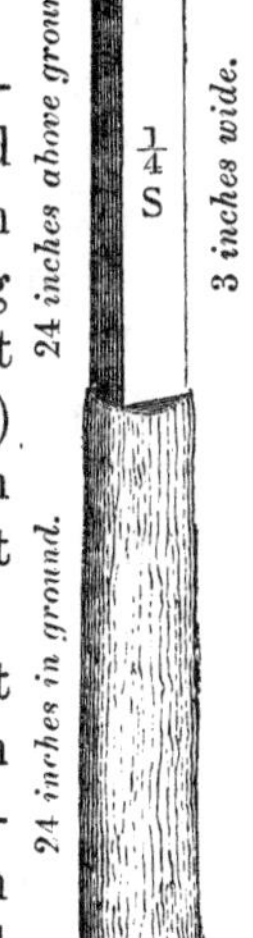

Quarter section corners are perpetuated in the following manner, to wit:

1. Quarter Section Posts.—Posts at quarter section corners must be 4 feet in length and 4 inches in diameter, and be planted or driven into the ground 2 feet; the part projecting being flattened or squared, so as to present a smooth surface 3 inches in width. (Fig. 10.)

The only mark required on a quarter section post is the characters "$\frac{1}{4}$ S." The corner must also be witnessed by two bearing trees.

2. Quarter Section Mounds.—Mounds at quarter section corners will be $4\frac{1}{2}$ feet in diameter at their base, and 2 feet in perpendicular height, the post being 4 feet in length and inserted in the ground 12 inches; it will also be marked and witnessed the same as the post without the mound.

3. Quarter Section Stones.—Stones used for quarter section corners must have the fraction "$\frac{1}{4}$" cut upon the *west* side of north and south lines, and on the *north* side of east and west lines, and must be witnessed by two bearing trees. (Fig. 11.)

Fig. 11.

4. A Tree, when found in place, should be marked and witnessed in the same manner as the post.

MEANDER CORNERS.

At the points where township or section lines intersect large ponds, lakes, bayous, or navigable rivers, posts are established at the time of running the lines, which are called *meander corners.* Either of the four modes described for perpetuating corners may be employed for meander corners.

1. MEANDER POSTS.—No marking is required on meander posts, but they must be witnessed by two bearing trees or pits. They should also be firmly inserted in the ground.

2. MEANDER MOUNDS.—The mound and post at meander corners should be of the same dimensions as those for the section and quarter section corners. The pit should be directly on the line, and 8 links further from the water than the mound. When the pit cannot be so located, its course and distance from the corner should be stated in the field book.

3. STONES or TREES may be employed to perpetuate meander corners, and when so used must be witnessed the same as meander posts.

FIELD BOOKS.

The field notes of the deputy surveyor are the official and permanent record of the boundaries of the public lands. They afford the elements from which the plats of the public surveys are constructed, and are the original and only source from which authentic descriptions of established boundaries can be obtained. It is of the highest importance that deputy surveyors should keep a faithful, distinct and minute record of everything officially done and observed by them or their assistants in their field operations. Carelessness or a want of strict fidelity on the part of the surveyor, will impair the value of his notes, if not indeed render them worse than worthless.

Deputy surveyors are especially enjoined to make themselves perfectly familiar with the requirements in this regard, and with the printed specimen field notes which accompany these instructions. They will also note particularly the following requirements:

1. Separate and distinct field books are required to be kept for the different kinds of lines surveyed; thus, there must be a separate field book for meridian and base lines, another for standard parallels or correction lines, another for exterior or township lines, and another for subdivisional or sectional lines.

2. The title-page of each field book will designate the kind of lines run, and describe the particular surveys, giving the name of the state or territory in a prominent line. State also by whom the survey was made, the number and date of the contract, and in separate lines the date of commencing and completing the work.

3. An index upon the plan illustrated in the specimen field book, referring to the page of each mile and to each kind of survey, must accompany the field notes.

4. The exhibition of every *mile* of surveying, whether of township or subdivisional lines, must be complete in itself, and be separated by a black line drawn across the page.

5. The notes should in all cases be taken precisely in the order in which the work is done on the ground, and must show all the perambulations, calculations, and field operations.

6. The descriptions of the surface, soil, timber, undergrowth, etc., on each mile of line run, should *follow* the notes thereof, and not be mixed up with them.

7. No abbreviations of words are allowable, except such as are constantly recurring, as "*sec.*" for section, "in. diam." for inches diameter, "chs." for chains, "lks." for links, "dist." for distance, "va." for variation, etc. For quarter section corner, "$\frac{1}{4}$ sec. cor." may be used, and for

14 inches long, 12 inches wide, and 3 inches thick, in describing a corner stone, "14 × 12 × 3," being particular to always preserve the same order of length, width, and thickness.

Proper names must never be abbreviated, however often their recurrence.

8. When surveys are commenced in one fiscal year and completed in another, the field book must be so kept as to show distinctly the amount of work done in each year separately. This requirement was adopted in 1863, and made imperative, in order that the General Land Office might be able to exhibit annually the amount of surveys actually executed in each fiscal year.

9. The notes must be written in precise and clear language, and the figures, letters, words, and meaning are always to be unmistakable. No leaf must be mutilated or obliterated, and none be taken out, whereby suspicion might be created that the missing leaf contained matter which the deputy believed it to be his interest to conceal.

OBJECTS AND DATA REQUIRED TO BE NOTED IN THE FIELD BOOKS.

1. The township and range, or a description of the particular locality of the operations, should precede the notes of the surveys, and be repeated at the top of each page.

2. Variation of the Needle.—The variation of the needle must always be stated in a separate line preceding the notes of measurement. At all points in the lines where any material *change* in the variation is found, such changes, with the exact points where they occur, must be carefully noted.

3. Courses and Measurements.—The course and exact length of every line run, noting all offsets therefrom, with the reasons and mode thereof.

4. Bearing Trees.—The kind and diameter of all bearing trees, with their courses and distances from their respective corners, and the precise relative position of *witness corners* to the true ones.

5. Mode Employed to Perpetuate the Corners.—State if it be a *post;* if a *tree* in place, give the name and diameter; if a *stone*, the kind and dimensions; if a *mound*, the material (earth or stone) of which it is constructed, the kind of memorial buried at the side of the post, the fact that it is erected in accordance with instructions, and the courses and distances of the pits from the center of the mound where necessity exists for deviating from the general rule.

6. Line Trees.—The name, diameter, and distance on line to all trees intersected.

7. Intersection of Land Objects.—The distances at which the lines intersect and leave any settlers' claim and improvement, prairie, river, creek, or other "bottom," swamp, marsh, grove and windfall; with the courses of the same at the points of intersecting and leaving them.

8. Intersection of Hills, etc.—The distances at which a line begins to ascend, reaches the top, begins to descend, and arrives at the foot of all remarkable hills and ridges, with their courses and *estimated* height in feet above the land of the surrounding country.

9. Intersection of Water Objects.—The distances on line to all rivers, creeks, and smaller streams of water, with their width at the points of intersection, and the course they bear; also, in the case of navigable waters, all the particulars of the mode by which the width is ascertained.

10. Bottom Lands.—Wet or dry, and if subject to inundation, to what depth. (See *Swamp Lands.*)

11. Lakes and Ponds.—Describe their banks and give their height; also the depth of water, and whether it be pure or stagnant.

12. Settlements and Improvements.—Towns and villages, Indian towns and wigwams, houses and cabins, fields, fences, and other improvements; groves, mill-seats, forges, and factories.

13. Springs.—Whether fresh, saline, or mineral, with the course of the streams flowing from them.

14. Minerals and Coal Beds.—Note all coal banks or beds, with a particular description of the same, as to quality, extent, and diggings therefor; and designate the localities *by the smallest legal subdivisions.* A recent law of Congress makes the strict observance of this requirement essential.

15. Roads and Trails.—Whence and whither, with their directions.

16. Rapids, Cascades, Cataracts, or falls of water, with the height of their fall in feet.

17. Precipices, Caves, Ravines, sink-holes, stone quarries, ledges of rocks, with the kind of stone they afford.

18. Natural Curiosities.—Interesting fossils, petrifactions, organic remains, etc.; also all ancient works of art, such as mounds, fortifications, embankments, ditches, etc.

19. Land Surface.—Whether level, broken, or hilly—1st, 2d, or 3d rate on each mile—1st rate to indicate extra quality, 2d rate good average, and 3d rate inferior quality.

20. Timber.—Name the several kinds of timber and undergrowth in the order in which they predominate, on each mile of line.

21. Dates.—State the month and day of the month in a separate line, immediately following the notes of each day's work.

22. General Description.—In subdivisional work the deputy must subjoin at the conclusion of the ordinary notes taken on line a *general description* of the township in the aggregate, as regards the face of the country, its soil and geological features, timber, minerals, water, etc.; and should add any further description or information touch-

ing any matter or thing connected with the survey which he may be able to afford and may deem useful or necessary to be known.

23. Names of Surveying Party.—The field book should also contain a list of the names of the persons employed in running, measuring, and marking the lines embraced in said notes, stating the respective capacities in which they severally acted.

24. Verification of Deputy Surveyor.—The deputy must append to each separate book of field notes his affidavit that all the lines described therein have been run and all corners established and perpetuated in strict conformity with instructions and the laws of the United States, and that the foregoing are the true and original field notes of such survey.

25. Verification of Assistants.—The chainmen, axman, and compassmen must also attest under oath that they assisted said deputy surveyor in executing said surveys, and that to the best of their knowledge and belief the work has in all particulars been performed strictly according to the instructions furnished by the Surveyor-General.

26. Approval and Certificate of Surveyor-General.—To each of the original field books the Surveyor-General will attach his official approval, and to the copies of the field notes transmitted to the General Land Office, he will affix his official certificate that they have been correctly copied from the originals on file in his office.

For a more full understanding of the manner of keeping the field book, and the forms requisite to be used, reference is made to the specimens which accompany these instructions. By giving due attention to these specimens and instructions, the surveyor will be enabled to fulfill all the requirements of law in this regard.

MEANDERING.

Large lakes, navigable rivers and bayous, are by law of Congress made public highways, and as the government surveys progress they are meandered and segregated from the public lands. Large ponds and water-courses not navigable are in some cases also meandered.

At those points where the lines of the public surveys intersect meanderable streams and bodies of water, "meander corners" are established at the time of running such lines. By the process of meandering, these corners are connected by ascertained courses and distances along the margin of the water, and the configuration of lakes, bayous, rivers, etc. is thereby obtained.

In meandering water-courses, where a distance is more than *ten chains* between stations, even chains only should be taken; but if the distance is *less* than ten chains, and it is found convenient to employ chains and links, the number of links should be a multiple of ten, thereby saving time and labor in testing the closing both in the field and in the Surveyor-General's office.

Standing with the face looking down stream, the bank on the right hand is termed the "right bank," and that on the left hand the "left bank." These terms are to be uniformly used to distinguish the two banks of a river or stream.

To meander a river the deputy will commence at a meander corner on the township line and proceed to course the sinuosities of the river bank, measuring the distance on each course, to the next meander corner on the same or another boundary of the township, entering the courses and distances in their proper places in the field book, and in the order in which they are taken, and noting the intersections with all intermediate meander corners.

1. Navigable Rivers.—All navigable rivers are to be

meandered on *both banks*, and care must be taken, in time of high water, not to mistake the margins of bayous or the borders of overflowed marshes or "bottoms" for the true river bank. Sufficient courses must be taken to follow closely the windings of the river and embrace all the land to its margin.

All streams of water emptying into the river, with the width at their mouth, the height of falls and cascades and the length of rapids, must be noted; and the banks, current, and bottom of the stream meandered, should be described in the field book.

2. RIVERS NOT NAVIGABLE.—Rivers not embraced in the class denominated "navigable" under the statute, but which are well-defined natural arteries of internal communication, and have a uniform width, will only be meandered *on one bank*. For the sake of uniformity, the surveyor will traverse the *right bank* when not impracticable; but where serious obstacles are met with, rendering it difficult to course along the right bank, he may cross to the left bank and continue the meanders as far as necessary; but all changes from one bank to the other will be made at the point of intersection of some line of the public surveys with the stream being meandered.

The subdividing deputies will be required to establish meander corners on both banks of such meanderable streams at the intersection of all section lines, and the distances across the river will be noted in the field book.

3. WIDE "FLATS."—Where wide, irregular expansions occur in rivers that are not navigable, and such expansions are permanent bodies of water, the area of which is more than forty acres, and embraces more than one-half of a legal subdivision of forty acres, they should be meandered on both banks.

4. ISLANDS IN RIVERS.—The precise relative position of islands in a township made fractional by the river in which the same are situated, must be determined trigo-

nometrically. Sighting to a flag or other object on the island from a special and carefully measured base line connected with the surveyed lines, and on or near the river bank, form connections between the meander corners on the river and points in direct line on the shores of the island, at which points establish meander corners, and calculate the distance across. The operation must be particularly and fully described in the field notes.

5. LAKES, PONDS, ETC.—Lakes embracing an area of less than *forty acres* will not be meandered. Long, narrow, or irregular lakes of larger extent, but which embrace *less than one-half* of the smallest legal subdivision, will not be meandered. Shallow lakes or bayous, likely in time to dry up or be greatly reduced by evaporation, drainage, or other cause, will not be meandered however extensive they may be, but such lakes should be described, and the facilities for draining or prospect of becoming dry from natural causes stated.

Posts will be established by the subdividing deputy, at the intersections of all the public lines with these lakes, the same as if they were to be meandered.

To meander a lake, pond, or bayou, commence at a meander corner on the township line, and proceed as directed in meandering a navigable river. Where the body of water lies entirely within the township boundaries, the deputy should commence at a meander corner established in subdividing, and from thence course around the entire pond or lake, noting the intersection with all the meander corners previously established.

To meander a pond lying entirely within the boundaries of a *section*, run and measure two lines to such pond from the nearest *opposite* section or quarter section corners, giving the courses thereof, and at each of the points where these lines intersect the margin of the pond, establish a "*witness point*" by fixing a post in the ground, and taking bearings and distances to adjacent trees, or if no trees are found, raising a mound.

The relative position of these points being thus definitely fixed in the section, commence at one of them and course to the other, noting the intersection, and thence to the place of beginning. The mode of proceeding must be fully set forth in the field book.

The meander notes must state particularly the corner from which they start, and the meanders of each fractional section are to be exhibited separately. All islands, rapids, and bars are to be noticed, and their exact situation indicated by intersections to their upper and lower points; also the head and mouth of all bayous.

The notes of meanders will be placed at the end of the notes of the township, and according to the dates when the work is performed, as illustrated in the specimen field notes. Following and composing a part of such notes will be given a description of the soil, timber, and depth to which the bottoms are subject to overflow.

The lakes, bayous, ponds, and so much of meanderable rivers as lie within the boundaries of a township are to be meandered at the time of subdividing the township, and the notes thereof will be annexed to and form a part of the field notes of such subdivisional survey.

No blazes or marks of any description are to be made on lines meandered between established corners.

SURVEY OF SWAMP LANDS.

By the act of Congress, approved September 28th, 1850, swamp and overflowed lands "unfit for cultivation" are granted to the state in which they are situated. These lands are selected and approved to the state according to the predominating character of the *smallest legal subdivision.* If the larger part of such subdivision is swamp and overflowed, it goes to the state; if otherwise, it is excluded from the grant, and retained by the government.

In order therefore to determine what lands fall to the

state under the swamp grant, it is necessary that the field notes of surveys, in addition to the other objects of topography required to be noted, should indicate the points at which the public lines enter and leave all lands coming within the purview of said grant. The deputy surveyor is charged with the responsible duty of describing with care and fidelity the true character of all lands within the field of his surveying operations, which may come under the denomination of "swamp and overflowed," or "unfit for cultivation."

The grant aforesaid does not embrace tracts subject to casual inundations, but only those where the overflow would wholly prevent the raising of crops without the aid of artificial means, such as levees, etc.; hence the deputy should state whether such lands are *continually and permanently* wet or subject to overflow so frequently as to render them totally unfit for cultivation, giving the depth of inundation as determined from indications on the trees, etc. The frequency of overflow should be set forth as accurately as possible, from a knowledge of the character of the stream which causes the same, and the general contour of the country contiguous, aided by such reliable information as may be obtained from persons acquainted with the facts.

The character of the timber, shrubs, plants, etc. growing on such tracts, and the contiguity of the premises to rivers, water-courses, or lakes should be stated.

The words "unfit for cultivation" are to be employed in addition to the usual phraseology in regard to entering or leaving such swamp, marshy, or overflowed lands. It may be that sometimes the margin of bottom, swamp, or marsh, in which such uncultivable land exists, is not identical with the margin of the body of land "unfit for cultivation;" and in such cases a separate entry must be made for each opposite the marginal distance at which they respectively occur.

But in cases where lands are overflowed by *artificial* means, such as dams for milling, logging, etc., such overflow will not be officially regarded, but the lines of the public surveys will be continued across the same, without setting meander posts, stating particularly in the notes the depth of the water and how the overflow was caused.

ORDER OF CARRYING FORWARD THE PUBLIC SURVEYS.

In the preceding pages we have explained—1st, the system of dividing the public lands into rectangular tracts of convenient size to be disposed of for pastoral and agricultural purposes; 2d, the manner of coursing, measuring, and marking the lines of the public surveys; 3d, the mode of establishing and perpetuating the corner boundaries of the public lands; and 4th, how the field book, which is the permanent record of everything officially done or observed by the deputy surveyor and his assistants, is to be kept, with full instructions as to the objects and data required to be recorded therein.

We are now prepared to explain the order in which the public surveys are carried forward in the field. First, then, let us suppose for illustration, that it is required to commence operations in a new territory where no public surveys have been made.

Meridians and Base Lines.—It will be remembered that all government surveys are projected from previously established meridians and base lines, or standard parallels. Before proceeding to lay off townships, therefore, a meridian and base line must be surveyed and marked, unless the proposed surveys are in continuation of those already executed in an adjoining state or territory, from an existing base.

Meridians may be run either north or south as may be necessary to reach the locality desired. They must, however, be run on a *due* north or *due* south course, the half mile, the mile, and the six mile corners being accurately measured and durably perpetuated according to the instructions in the preceding pages. *Base lines* may be run on either a *due* east or *due* west course from the meridian, planting the half mile, mile, and township corners at the prescribed intervals, FULL MEASURE.

Initial Point.—The first step in proceeding to establish a meridian and a base line, will be to select some prominent natural land-mark convenient to the locality where the earliest surveys will be needed, for an initial or starting-point. An isolated, well-defined mountain, or the point of confluence of two rivers, afford favorable objects for the purpose indicated. If these are not to be found, some other permanent natural object should be sought for.

Standard Parallels or Correction Lines.—It may also be necessary to run one or more standard lines in the early part of the surveying operations, for it will be borne in mind that *all range lines* are to be run *north* from base or standard lines. Standard parallels, like base lines, may be run on either a due east or due west course from the meridian, planting the half mile, mile, and township corners at the prescribed intervals, *full measure.*

These lines form the skeleton or framework upon which to hang or build up the public surveys. Having provided the proper basis for this operation as directed, the deputy surveyor will next proceed to survey—

Principal Meridians.—As the public surveys progressed westward, six meridianal lines, denominated Principal Meridians, have been established and designated in numerical order from east to west, to wit:

The 1st Principal Meridian runs north from the mouth of the Great Miami river, between the States of Ohio and Indiana to the south boundary of Michigan.

The 2d Principal Meridian runs north from the mouth of the Little Blue river, through the center of the State of Indiana to its northern boundary.

1. EXTERIOR OR TOWNSHIP LINES.

Systematic order is observed in running township lines, and the perambulations of the deputy surveyor are fully illustrated by Diagram A, on page 54.

TOWNSHIPS WEST OF THE MERIDIAN.—Begin at the first pre-established township corner on the base line, west of the meridian, which will be the southwest corner of Township 1 N., Range 1 W., marked No. 1 on the diagram: thence north on a true meridian line 480 chains, establishing the section and quarter section corners thereon as per instructions, to No. 2, where establish the corner to Townships 1 and 2 N., Ranges 1 and 2 W.; thence east on a *random* line, setting *temporary* section and quarter section stakes to No. 3, where measure and note the distance at which the eastern boundary is intersected north or south of the true or established corner. Then, calculating a

---

The 3d Principal. Meridian runs north from the mouth of the Ohio river through the center of the State of Illinois, terminating at its northern boundary.

The 4th Principal Meridian runs north from the Illinois river through the western part of Illinois and the center of Wisconsin to Lake Superior.

The 5th Principal Meridian runs north from the confluence of the Arkansas and Mississippi rivers through the eastern portion of the States of Arkansas, Missouri, and Iowa, and governs the surveys in Minnesota lying west of the Mississippi river, and also in Dakota lying east of the Missouri river.

The 6th Principal Meridian commences on the Arkansas river, in the State of Kansas, and runs north through the eastern part of Kansas and Nebraska, terminating at the Missouri river.

Independent Meridians.—In addition to the six principal meridians above described, a number of Independent Meridians have been established in the newer states.

In New Mexico the surveys are reckoned from the Independent Meridian of New Mexico. In Utah, the Independent Meridian is styled the "*Salt Lake Meridian.*" The surveys in Oregon and Washington Territory are governed by an Independent Meridian called the "*Willamette Meridian;*" and in California there are three Independent Meridians governing the different surveys in that state, named respectively "*Humboldt Meridian,*" "*Mt. Diablo Meridian,*" and "*St. Bernardino Meridian.*" The surveys in Nevada are numbered from the "Mt. Diablo Meridian" in California.

DIAGRAM A.

1st COR. LINE N.
SILVER LAKE
MERIDIAN
Placer City
Clear R.
Deposit
Salmon River
BASE LINE
BEAVER RIVER
Valley City
1st CORRECTION LINE SOUTH

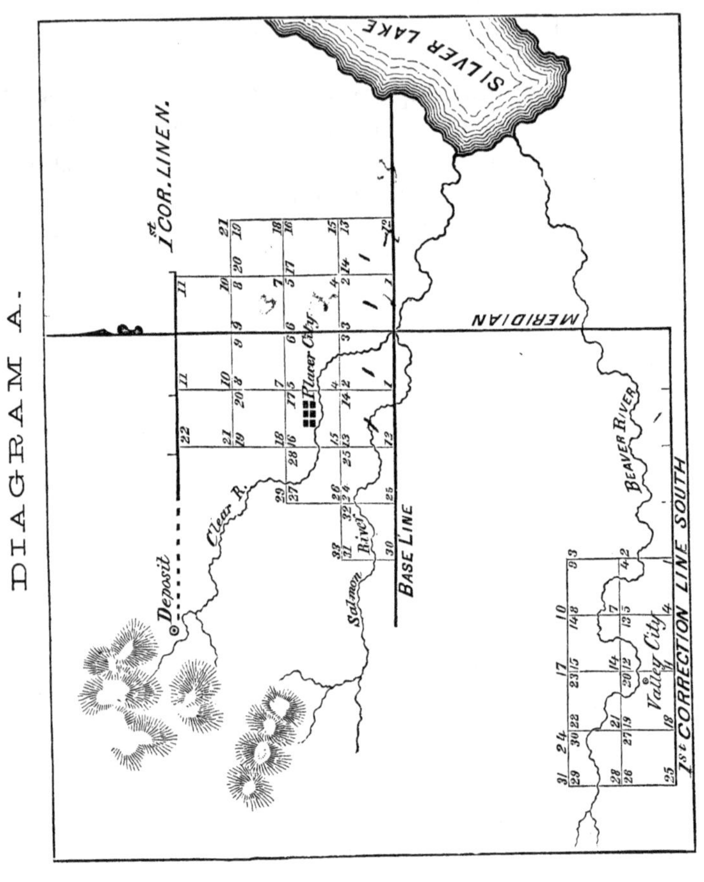

course that will run a *true* line back to the corner from which the random started, run and measure westward, to No. 4, which is identical with No. 2, establishing the *permanent* corners on said line and obliterating the temporary corners on the random, throwing the excess or deficiency in measurement on the west end of the line.

If in running the above random line the deputy should fall short or overrun in length, or intersect the township boundary at more than 3 *chains and* 50 *links* north or south of the true corner, either of which would indicate a material error in the work, the lines must be *retraced*, even if found necessary to remeasure the meridianal boundaries of the township.

Proceed in the same manner from No. 4 to No. 5—from No. 5 to No. 6—from No. 6 to No. 7—and so on to No. 10. From No. 10 run north, still on a true meridian, establishing the mile and half mile corners until the standard parallel is reached at No. 11, throwing the excess over or deficiency under 480 chains upon *the last half mile*, according to law; and at the intersection establish the township "closing corner," measuring and noting the distance to the nearest "standard corner" on said parallel.

Should it happen from any cause that the extension of the correction parallel above the field of operations has been delayed, the deputy will plant the corner of the township in place, subject to correction when the parallel shall be extended.

The deputy will then return to the S. W. corner of Township 1 N , Range 2 W. on the base line, at No. 12, and proceed in a similar manner to No. 33, returning and running each range line north from the base line.

It is sometimes desired to reach a locality not intersected by the meridian or base line. The manner of accomplishing this is illustrated by Diagram A. For example, the meridian may be extended south to such distance as may be necessary to run a correction parallel therefrom to serve as a base for the desired surveys. The manner of surveying the township exteriors from said parallel is similar to that already explained and indicated by the numbers on the diagram.

Townships East of the Meridian.—Begin at No. 1, Diagram A, being the southeast corner of Township 1 N., Range 1 E., and proceed as with the townships west of the meridian, except that the random lines will be run and measured *west* and the *true* lines *east*, throwing the excess over or deficiency under 480 chains on the *west end* of the line. The deputy will therefore commence his measurement with the length of the excessive or deficient half section boundary on the west of the township, and then the remaining measurements will all be exact miles and half miles.

With the notes of the exterior lines of townships, the deputy is required to submit a plat of the lines run, on a scale of 2 inches to the mile, on which are to be noted all the objects of topography on line necessary to illustrate the notes, to wit: The distances on line at the crossing of streams, so far as such can be noted on the paper, and the direction of each indicated by an arrow-head pointing down stream; also the intersection of lines by prairies, marshes, ponds, swamps, ravines, lakes, hills, mountains, and all other matters indicated by the notes, to the fullest extent practicable.

DIAGRAM B.

SUBDIVISION OF A TOWNSHIP INTO SECTIONS.

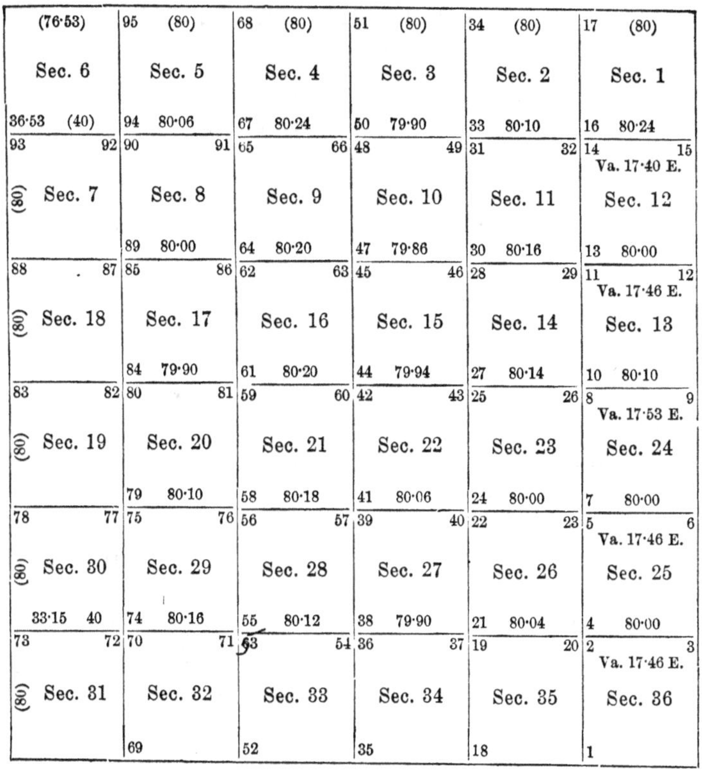

(76·53) Sec. 6 36·53 (40)	95 (80) Sec. 5 94 80·06	68 (80) Sec. 4 67 80·24	51 (80) Sec. 3 50 79·90	34 (80) Sec. 2 33 80·10	17 (80) Sec. 1 16 80·24
93 92 (80) Sec. 7	90 91 Sec. 8 89 80·00	65 66 Sec. 9 64 80·20	48 49 Sec. 10 47 79·86	31 32 Sec. 11 30 80·16	14 15 Va. 17·40 E. Sec. 12 13 80·00
88 87 (80) Sec. 18	85 86 Sec. 17 84 79·90	62 63 Sec. 16 61 80·20	45 46 Sec. 15 44 79·94	28 29 Sec. 14 27 80·14	11 12 Va. 17·46 E. Sec. 13 10 80·10
83 82 (80) Sec. 19	80 81 Sec. 20 79 80·10	59 60 Sec. 21 58 80·18	42 43 Sec. 22 41 80·06	25 26 Sec. 23 24 80·00	8 9 Va. 17·53 E. Sec. 24 7 80·00
78 77 (80) Sec. 30 33·15 40	75 76 Sec. 29 74 80·16	56 57 Sec. 28 55 80·12	39 40 Sec. 27 38 79·90	22 23 Sec. 26 21 80·04	5 6 Va. 17·46 E. Sec. 25 4 80·00
73 72 (80) Sec. 31	70 71 Sec. 32 69	53 54 Sec. 33 52	36 37 Sec. 34 35	19 20 Sec. 35 18	2 3 Va. 17·46 E. Sec. 36 1

FORMER MODE OF NUMBERING LOTS.

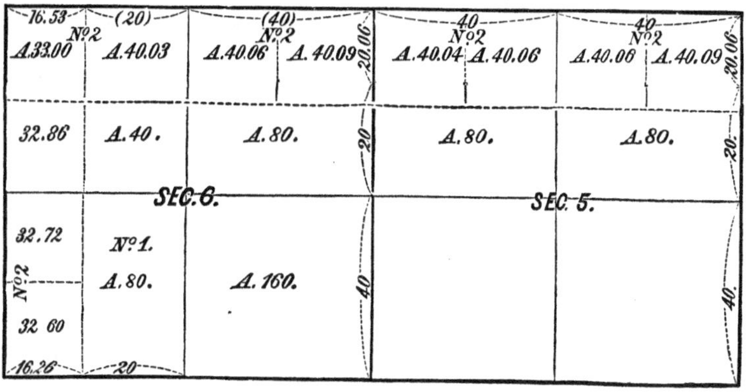

16.53 (20) (40) 40 40
No. 2 No. 2 No. 2 No. 2
A. 33.00 A. 40.03 A. 40.06 A. 40.09 20.06 A. 40.04 A. 40.06 A. 40.06 A. 40.09 20.06
32.86 A. 40. A. 80. 20 A. 80. A. 80. 20.
SEC. 6. SEC. 5.
32.72 No. 1. A. 80. A. 160. 40 40.
No. 2
32 60
16.26 20

2. SUBDIVISION OR SECTION LINES.

The township lines being run, the next step will be to subdivide or sectionize the township by surveying it into sections. By a rule of the department a deputy surveyor is not permitted to subdivide townships the exteriors of which were run by himself. The reason for this regulation is that one deputy may be a check upon the doings of another, thereby detecting imperfect or fraudulent work and securing accuracy in the execution of the public surveys.

Before proceeding to subdivide a township, the deputy must ascertain what change, if any, has taken place in the magnetic variation as it existed at the time the exteriors were run, and also compare his chaining with the original measurements. For this purpose he is required to retrace the first mile, both of the south and east boundaries of each township, and any discrepancy either in the variation or chaining must be noted in the field book.

Having adjusted the compass to a variation which will retrace the eastern boundary of the township, the deputy will begin at the first mile corner west on the south boundary, which will be common to sections 35 and 36—see No. 1, Diagram B—thence due north 40 chains, where establish quarter section corner, and continue due north 80 chains, to No. 2, at which point plant corner to sections 25, 26, 35, and 36.

From No. 2, run, on *random line*, without blazing, *due east*, setting temporary quarter section post at 40 chains, and continuing on to the east boundary at No. 3. If the township line is intersected exactly at the section corner thereon, the random may be blazed back and established as the true line; but if said random strike the boundary either north or south of the section corner, the distance of the point of intersection from said corner must be measured and noted, and a course calculated that will run a true line from the section line on the boundary back to No. 4, from which the random started. The permanent quarter section corner must be established on the true line at a point equidistant from the two section corners, according to the requirements of law, and the temporary post on the random should be pulled up.

From No. 4 proceed due north between sections 25 and 26, as before, to No. 5, where establish corner to sections 23, 24, 25, and 26; thence east on a random line to No. 6, correcting back, as before directed, to No. 7.

Proceed in this manner to survey the lines of the first tier of sections up to No. 16, which is the corner to sections 1, 2, 11, and 12. From this corner run due north on a *random* between sections 1 and 2 to No. 17, the north boundary of the township. If the random does not close exactly on the section corner pre-established, the distance of the intersection from said corner must be measured and noted, and a course calculated that will run a *true* line south to No. 16, from which the random started, the same as randoms east, except that the permanent quarter section corner must be planted exactly *forty chains* from the interior section corner (No. 16), thereby throwing the excess or deficiency in measurement on the last half mile, according to law.

When the township boundary is a base line or standard parallel, no random is required, but a *true* line must be run due north, establishing the permanent quarter section corner at just 40 chains, and at the point of intersection with said parallel or base line erect a "closing corner," carefully measuring and noting the distance to the corresponding "standard corner."

The first tier of sections being thus completed, the deputy will return to No. 18 on the south boundary, and proceed in the same manner to survey the second tier, closing on the interior section corners just established, the same as he did upon those on the east boundary of the township.

In surveying the fifth section line between the fifth and sixth tiers of sections, not only an east random is run between the sections, but a random must also be run due west to the range line, and corrected back the same as between the sections in the first tier, except that the permanent quarter section corner must be established exactly *forty chains* from the interior section corner, as required on the north boundary, throwing the excess or deficiency of measurement upon the last half mile or outside quarter section.

With his instructions for making subdivisional surveys, the Surveyor-General will furuish the deputy with a diagram of the exterior lines of the townships he is to subdivide, on a scale of two inches to the mile, upon which are laid down the measurements of each mile on said boundaries, the magnetic variations of each mile and the particular description of each corner, the letters "P. M." being used to signify "post in mound."

On this diagram the subdividing deputy will make appropriate sketches of the various objects of topography as they occur on his lines, showing not only the points at which they occur, but also the direction and position of each between the lines, so that the objects will be complete and properly connected in the showing.

The perambulations of the deputy surveyor in subdividing a township are fully shown by Diagram B, and the mode and order described are to be followed in all cases.

LAWS, REGULATIONS, AND INSTRUCTIONS RELATING TO DEPUTY SURVEYORS.

The public lands are surveyed by deputy surveyors under contracts made with the Surveyors-General in the respective surveying districts.

The first section of the act of May 30th, 1862, provides

that contracts for surveying the public lands shall not become binding upon the United States until approved by the Commissioner of the General Land Office, except in such cases as the Commissioner shall otherwise especially order; but it should be distinctly understood that the General Land Office does not assume to dictate to whom surveying contracts shall be let, nor to exercise any control whatever in the matter, so long as they are let to competent and faithful men. The giving out of these contracts is left entirely with the Surveyors-General, and to them alone should applications for employment as government surveyors be made.

The position of deputy surveyor is open to all competent surveyors from any part of the United States who are loyal men. They must be men of integrity and well skilled in their profession, and the department will not knowingly approve a contract made with any party who does not possess these qualifications.

To United States deputy surveyors is intrusted the duty of establishing permanent boundaries of the public lands, and it is of the highest importance that their work be accurately and faithfully performed. The surveys are usually carried forward in advance of settlement, and are frequently prosecuted in remote sections of the country, far away from any inhabitants; but although thus removed from the observant eye of the government, the honest deputy will feel none the less his obligations to execute his work with the strictest fidelity.

Bad surveying and unfaithfulness in the erection and marking of corners are sure to be exposed sooner or later, and when they are detected the reputation of the deputy is gone forever, and he is liable to punishment for perjury and fraud.

The responsibility under which deputy surveyors act is defined by the second section of an act of Congress, approved Aug. 8th, 1846, entitled "An act to equalize the compensation of the Surveyors-General of the public lands of the United States, and for other purposes," as follows:

"SECT. 2. That the Surveyors-General of the public lands of the United States, in addition to the oath now authorized by law to be administered to deputies on their appointment to office, shall require each of their deputies, on the return of his surveys, to take and subscribe an oath or affirmation that those surveys have been faithfully and correctly executed according to law and the instructions of the Surveyor-General; and on satisfactory evidence being presented to any court of competent jurisdiction, that such surveys, or any part thereof, had not been thus executed, the deputy making such false oath or affirmation shall be deemed guilty of perjury, and shall suffer all the pains and penalties attached to that offense; and the District Attorney of the United States for the time being, in whose district any such false, erroneous, or fraudulent surveys shall have been executed, shall, upon the application of the proper Surveyor-General, immediately institute suit upon the bond of such deputy; and the institution of such suit shall act as a lien upon any property owned or held by such deputy, or his sureties, at the time such suit was instituted."

Every consideration of duty and of interest prompts the deputy to faithfulness and fidelity in the execution of the work confided to him. The loss of honor and of a good name, a loss which can never be regained in the public service, are the forfeit of a failure in this regard, and the history of all past experience in this branch of the public service demonstrates that those deputies and only those who have thus shown themselves worthy recipients of public trusts, have ever been successful or prosperous.

CONTRACT AND BOND.

Before entering upon the survey, the deputy is required to execute a contract and bond with the Surveyor-General in the following form, to wit:

This Agreement, made this (*twenty-fourth*) day of (*May*), 1867, between (*Solomon Sharp*), Surveyor-General of the United States for (*the Territory of Idaho*), acting for and in behalf of the United States, of the one part, and (*Peter Traverse*), deputy surveyor, of the other part, WITNESSETH, That the said (*Peter Traverse*), for and in consideration of the conditions, terms, provisions, and covenants hereinafter expressed, and according to the true intent and meaning thereof, doth hereby covenant and agree with the said (*Solomon Sharp*), in his capacity aforesaid, that (*he*) the said (*Peter Traverse*) in (*his*) own proper person, with the assistance of such chainmen, axmen, and flagbearers as may be necessary, agreeably with the laws of the United States, and in strict conformity with the printed Pamphlet and Manual of Surveying Instructions, issued by the General Land Office, which are hereby incorporated with and made a part of this contract, and with such special instructions as (*he*) may receive from the Surveyor-General in conformity therewith, will well, truly, and faithfully (*survey the exterior lines of the following described Townships, to wit: Townships* 2, 3, 4, *and* 5 *N. Ranges* 6, 7, *and* 8 *E. of the Meridian, in the Territory of Idaho*), and that (*he*) will complete these surveys in the manner aforesaid, and return the true and original field notes thereof to the office of the said Surveyor-General on or before the (*twentieth*) day of (*November*) next ensuing the date hereof,—acts of God excepted,—on penalty of forfeiture, and paying to the United States the sum mentioned in the annexed bond, if default be made in any of the foregoing conditions. And it is further expressly stipulated and made a condition of this contract, that the surveys herein described shall not be commenced before the first day of the fiscal year ending the 30th day of June, 1868.

And the said (*Solomon Sharp*) in his capacity aforesaid. covenants and agrees with the said (*Peter Traverse*) on account of the United States, that there shall be paid him by the Treasury Department, upon the receipt of (*his*) account at the General Land Office, properly certified by said (*Solomon Sharp*) in his capacity aforesaid, and accompanied by the approved plats of the surveys for which the account is rendered, as a full compensation for the whole expense of surveying and making return thereof (*eight dollars*) per mile, for every mile and part of mile actually run and marked in the field, offsets and random lines not included.

And it is further understood and agreed, between the parties to this agreement, that the said surveys will not be approved by the said (*Solomon Sharp*) in his capacity aforesaid, unless they shall be found to be in exact accordance with the requirements in the printed Pamphlet and Manual of Surveying Instructions: *Provided* also, No member of Congress or subcontractor shall have any part in this contract, and that no payment shall be made for any surveys not executed by the said deputy surveyor in (*his*) own proper person.

IN TESTIMONY WHEREOF, The parties to these articles of agreement have hereunto set their hands and seals the day and year first above written.

SOLOMON SHARP, Surveyor-General. [L. S.]
PETER TRAVERSE, Deputy Surveyor. [L. S.]

Signed, sealed, and acknowledged
before us,
JOHN SMITH,
THOMAS JONES.

I, Peter Traverse, deputy surveyor, do solemnly swear (*or affirm*), that *I* will faithfully and impartially execute the surveys mentioned in the foregoing contract to the best of (*my*) skill and ability.

PETER TRAVERSE, Deputy Surveyor.

Sworn to and subscribed before me,
at (*Golden City*), in the (*Territory of Idaho*), this (24*th day of May*), 1867.

WALTER CARROLL, Notary Public.

KNOW ALL MEN BY THESE PRESENTS, That we (*Peter Traverse*), deputy surveyor, as principal, and (*John Smith*) and (*Thomas Jones*), as sureties, are held and firmly bound unto the United States in the sum of (*two thousand*) dollars, lawful money of the United States (being double the estimated amount which would be due by the United States to the said [*Peter Traverse*], on the completion of the surveys named in the foregoing contract), for which payment, well and truly to be made, we bind ourselves, our heirs, executors, and administrators, and each and every of us and them, jointly and severally, firmly by these presents; signed with our hand. and sealed with our seals, this (*twenty-fourth*) day of (*May*), 1867.

THE CONDITION OF THE ABOVE OBLIGATION IS SUCH, That if the above bounden (*Peter Traverse*), deputy surveyor, shall well and truly and faithfully, according to the laws of the United States and the instructions of the said Surveyor-General, and in strict conformity with the printed Manual of Surveying Instructions, make and execute the surveys which are required of (*him*) to be made by the foregoing contract, and return the field notes of the said surveys to the Surveyor-General, in the manner and within the period named in the said contract, then this obligation to be void, or otherwise it shall remain in full force and virtue.

PETER TRAVERSE,
Deputy Surveyor.

Signed, sealed, and acknowledged
before us,

JAMES HILL,
Golden City, Idaho T.
HENRY STEVENS,
Boise City, Idaho T.

I (*William Brown*) do certify that in my opinion the sureties to the above bond are sufficient, and I hereby approve the same.

Witness my hand and seal at (*Golden City, I. T.*), this (*24th*) day of (*May*), 1867.

[L. S.] WILLIAM BROWN.

OATH

Prescribed to be taken by all Persons in the Public Service, by Act of Congress, Approved July 2d, 1862.

I (*Peter Traverse, deputy surveyor*) do solemnly (*swear*) that I have never voluntarily borne arms against the United States since I have been a citizen thereof, that I have voluntarily given no aid, countenance, counsel, or encouragement to persons engaged in armed hostility thereto; that I have neither sought nor accepted nor attempted to exercise the functions of any office whatever under any authority or pretended authority in hostility to the United States; that I have not yielded a voluntary support to any pretended government, authority, power, or constitution within the United States, hostile or inimical thereto. And I do further (*swear*) that, to the best of my knowledge and ability, I will support and defend the Constitution of the United States against all enemies, foreign and domestic; that I will bear true faith and allegiance to the same; that I take this obligation freely, without any mental reservation or purpose of evasion; and that I will well and faithfully discharge the duties of the office on which I am about to enter. So help me God.

PETER TRAVERSE, [L. S.]
Deputy Surveyor.

} *ss.*

Sworn and subscribed to before me, this (*twenty-fourth*) day of (*May*), 1867. WALTER CARROLL, Notary Public.

Deputy surveyors are required to verify by their oath that the surveys embraced in their contracts have been executed in strict conformity with instructions, the requirements of the surveying Manual, and the laws of the United States. The deputy cannot consistently make this oath if the work is done by separate parties in other parts of the field from that in which he is himself operating; therefore, by a rule of the department, deputy surveyors are required to execute the surveys named in their contract "*in their own proper persons*," and to make oath that of their own personal knowledge the work is done in accordance with instructions and the laws of the United States, and no surveys not so executed and attested will be paid for.

It will be seen, therefore, that a deputy surveyor can take but *one surveying party* into the field at one time, and that party must be under his own supervision *in person.*

When one deputy does the work under a joint contract, he may verify the same.—When two deputies enter into joint contract for certain surveys, and *only one* of them goes into the field, if that one, with a single surveying party, executes all the work in person, his affidavit alone as surveyor, attached to the field notes, will be deemed sufficient, and no impediment to the payment of his account will result therefrom.

Two active deputies under a joint contract each to verify his own work.—If two deputies, joint parties in a contract, *both* go into the field, each with a separate surveying party, the field notes must show clearly the particular surveying done by each deputy. The date and the name of the deputy will be stated at the beginning and end of the notes of every continuous part of such survey executed by him, so that it may be distinctly seen by whom each mile of line was run.

The following form of affidavit is prescribed, to be attached to the field notes in cases of joint surveys, to wit:

FORM OF AFFIDAVIT FOR JOINT SURVEYS.—"I, A. B., deputy surveyor, do solemnly swear that, in pursuance of a joint contract, wherein A. B. and C. D. are joint contractors with S. G., United States Surveyor-General for , bearing date the day of , 18 , I have well, faithfully, and truly, in my own proper person, and in strict conformity with the instructions furnished by the Surveyor-General, the surveying Manual, and the laws of the United States, surveyed all those parts or portions of as are represented in the foregoing field notes as having been surveyed under my direction; and I do further solemnly swear that all the corners of said surveys have been established and perpetuated in strict accordance with the surveying Manual and printed instructions, and that the foregoing are the true and original field notes of such survey."

The separate affidavit of each deputy, in the above form, will be attached to the field notes of joint surveys.

Operations in the Field when to commence.—The surveys cannot be commenced in advance of the year for which the means is provided by Congress, and no moneys can be used to pay for work done before they were appropriated. This is an invariable rule, to be rigidly observed.

The object of this restriction is to keep back the surveying operations to the legitimate period of time contemplated in the appropriations. These appropriations are made with reference to the current necessities of given years, and if allowed to be absorbed in advance, the purposes of Congress in providing stated sums annually to carry forward the public surveys would be defeated.

[It would be much better both for the deputies and for the government if Congress would change the present custom of making the surveying appropriations at the last of the session. By this system a large portion of the best part of the year is passed before the appropriations are made. The deputies are thereby prevented from going early into the field, and the season for active operations is greatly curtailed. The natural tendency of this system is to cause an undue pressure of work as the close of the season approaches.]

In order, however, to enable deputy surveyors to avail themselves of as much of the season belonging to the fiscal year as possible, the Surveyor-General is notified by mail or telegraph, as circumstances may determine, when the appropriations are passed, but no surveying chargeable to such appropriation must be done before receiving such notice.

THE SURVEYING MANUAL AND INSTRUCTIONS OF THE COMMISSIONER ARE MADE A PART OF THE SURVEYING CONTRACTS BY LAW.

By the 2d section of the act of Congress entitled "An act to reduce the expenses of the survey and sale of the public lands in the United States," approved May 30th, 1862, it is provided: "That the printed Manual of Instructions relating to the public surveys, prepared at the General Land Office, and bearing date February twenty-second, eighteen hundred and fifty-five, the instructions of the Commissioner of the General Land Office, and the special instructions of the Surveyor-General, when not in conflict with said printed Manual or the instructions of said Commissioner, *shall be taken and deemed a part of every contract for surveying the public lands of the United States.*"

CONTRACTS MUST BE APPROVED BY THE COMMISSIONER.

The 1st section of the act of May 30th, 1862, provides that contracts for the survey of the public lands shall not become binding upon the United States until approved by the Commissioner of the General Land Office, except in such cases as the Commissioner shall otherwise especially order.

In the more remote districts, as for instance California, Oregon, and Washington Territory, the Commissioner's approval is given in advance, with certain limitations, to avoid impracticable delays.

PRESCRIBED LIMITS FOR CLOSINGS AND LENGTH OF LINES IN CERTAIN CASES.

1. Every north-and-south section line, except those terminating in the north boundary of the township, must be 80 *chains* in length.

2. The east-and-west *section lines*, except those terminating in the west boundary of the township, are to be within 100 *links* of the actual distance established on the south boundary line of the township for the width of said tier of sections.

3. The north boundary and south boundary of any one section, except in the extreme western tier, are to be within 100 *links* of equal length.

4. The meanders within each fractional section, or between any two meander posts, or of a pond or island in the interior of a section, must close within 1 chain and 50 links.

5. In running *random* township exteriors, if such random lines fall short or overrun in length, or intersect the eastern or western boundary, as the case may be, of the township, at more than 3 *chains and* 50 *links* north or south of the true corner, the lines must be *retraced*, even if found necessary to remeasure the meridianal boundaries of the township.

MEASURING DISTANCES OVER LAKES AND RIVERS.*

It frequently happens in surveying the public lands that lakes, rivers, and bayous, the distance across which cannot be measured with the chain, interrupt the public lines. The following illustrations will assist the inex-

FIGS. 1, 2 AND 3.

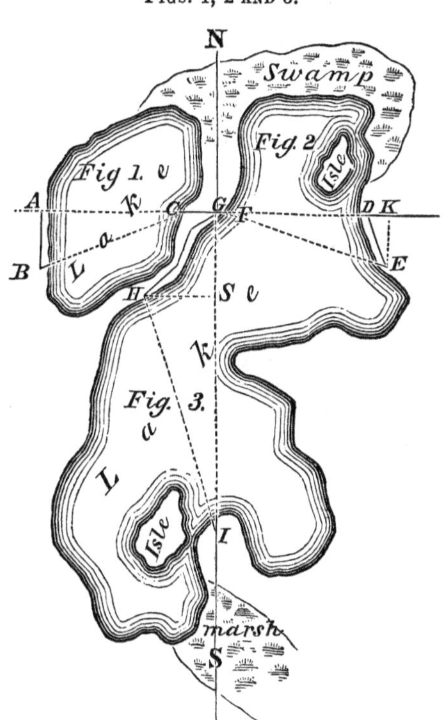

perienced surveyor in passing such obstacles. They are given on the principle of reducing the base, whatever

* The four following illustrations are taken from Burt's "Key to the Solar Compass." Mr. Burt was the inventor of this instrument, which is a great improvement on the magnetic compass, and has been adopted by the government to be used in the running of standard and township lines.

may be its course or courses, to a right-angled base to the course of the line to be measured. This can be readily done if care be taken to run and measure the base, at such angles that their latitude and departure can be taken from the traverse table.

FIGURE 1.

Distance required over lake from A to C, course east—right-angled base—from A to B 690 links. Angle at C 20° 20.′

Natural co-tangent of the angle at C .	= 2·698525
Multiplied by base A B	690
	242867250
	16191150
	1861·982250

Over lake 1862 links.

FIGURE 2.

Distance required over lake from D to F, course west—from D to E, S. 20° E. 752 links—gives 707 links southing, which is the right-angled base K G, and 257 links easting from G to K. Angle at F 15¾°.

Natural co-tangent of the angle at F . .	= 3·545732
Multiplied by the base K E	707
(Nat. co-tan. F × K E)—K D = D F.	24820124
(3·545732 × 707) = 2507 — 257 = 2250.	24820124
	2506·832524
Subtract distance from D to K.	257
Distance from D to F . . .	2250 lks. nearly.

FIGURE 3.

Distance required over lake from G to I, course south. To obtain a base in this example, we run

	Southing.	Westing.
S. $55\frac{3}{4}$ degrees, W. 400 . .	225	331
S. $19\frac{1}{4}$ do. W. 440 . .	415	145
S. 50 do. W. 548 . .	352	420
Distance from G to S . .	992 . . H to S	896 links.

Co-tangent of the angle 16° 38′ at I .	3·347319
Multiply by H S	896
	20083914
	30125871
	26778552
Distance from S to I	2999·197824
Add distance from G to S	992
Distance over lake from G to I . . .	3991 links.

Deputy surveyors will be allowed pay for the distance across lakes or ponds not meandered where they are required to continue the lines of the public surveys across them; but no offsets or lines run in triangulating will be paid for.

Where the distance across a lake or other body of water is ascertained by offsetting, it is not enough to say in the field notes "8·65 over lake and set a meander corner," but the *mode* by which the distance is ascertained must be stated and described in full.

DISTANCE OVER A RIVER BY "OFFSET."

Example.—Fig. 4.

In running a line north, intersect the right bank of a river at A (course N.N.E.) and erect an object, turn the compass sights to west, to an object at B, and pass over the river to it, then run and measure a line north to C, and "offset" east into line at D, the distance between A and D will be equal to the distance between B and C. Or, if a line be run and measured from A, N. 60° E., until an object in line at D bears N. 30° W., the distance A D will be twice that of A E, for the reason that the triangle thus formed is one-half of an equilateral triangle.

Fig. 4

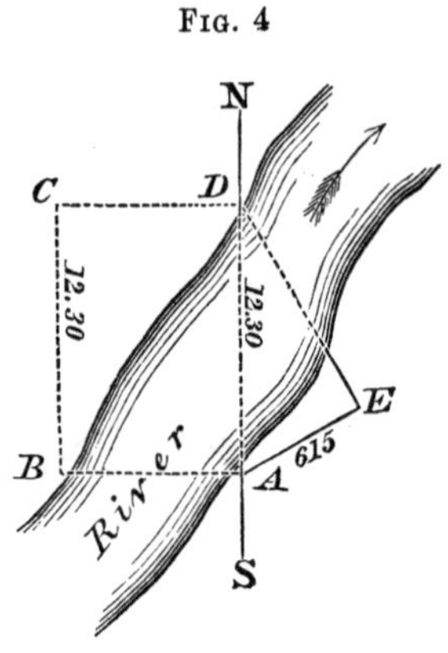

Frequently offsets are made in passing small lakes, bends of rivers, etc.: sometimes the distances can be advantageously taken over such obstacles with the telescope and rod. Also, it often happens that a suitable angle can be taken, and the base to that angle measured afterwards; in such cases the distance can be taken from the traverse table; but if no traverse, or other proper tables are at hand, the following angles, on a right angle base, and the multiplier to it, will give the distance. These may be committed to memory.

Angle 11°, 18′, multiply the base by 5.
" 14, 2, multiply the base by 4.
" 18, 26, multiply the base by 3.
" 21, 41, multiply the base by 2·5.
" 26, 34, multiply the base by 2.

SHORT METHOD OF FINDING THE AREA OF A MULTANGULAR FIELD.

Example, showing how to reduce the plot of a multangular field to a field of equal area having only three or four sides, by which its contents may be readily found.

To reduce such a field the only instruments required, after the meanders are properly laid down, are a good parallel-rule* and a fine protracting point.

In the following figure, first extend the base E H to an

* The triangle and the rule are the best.

indefinite length; then placing the rule on the angles 1 and 3, move it parallel from the angles 1 and 3 to the

Fig. 5.

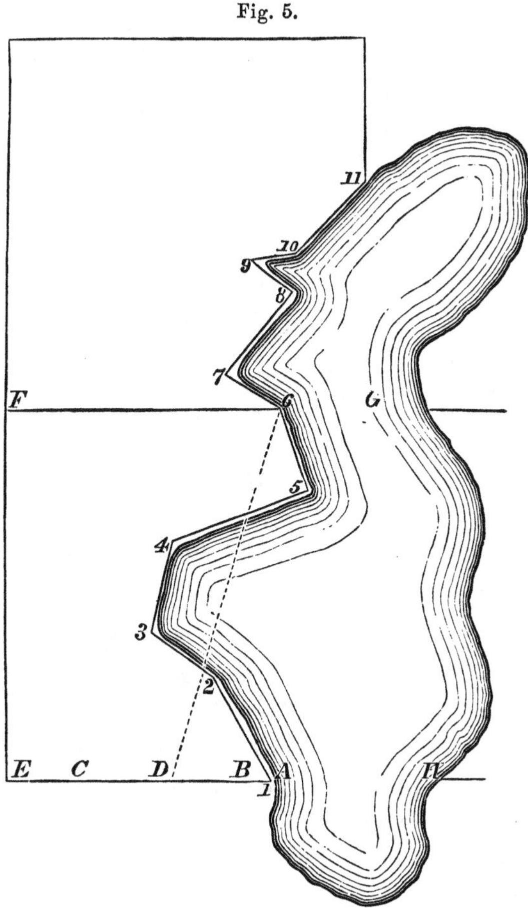

angle 2, and mark the exact point of intersection at A, on the base E H. Now place the rule on A and the angle 4; then move it parallel to the angle 3, finding the point B on the base E H; place the rule on B and the angle 5, and move, parallel, to the angle 4, finding the point C on the base E H. Now place the rule on the point C, and the terminating point 6 on the line F G, and move the rule, parallel, to the angle 5, finding the point D on the base E H, from which point draw a line to 6, the process then being complete. The line D 6 thus drawn, leaves the same area of lake to the left, that there is of land to the right. (Fig. 5.)

Any figure may be calculated upon the same principle by drawing a base and erecting a perpendicular line from it, passing through the figure. Place the rule at *a* and *c*, then move, parallel, back to *b*, marking the point 1 on the base; then from 1 to *d*, and move forward to *c* and so on to the angle at *i*, leaving a triangle to the right of the perpendicular. Proceed in like manner with that portion of the figure to the left of the perpendicular line, throwing it into two triangles. (Fig. 6.)

Fig. 6.

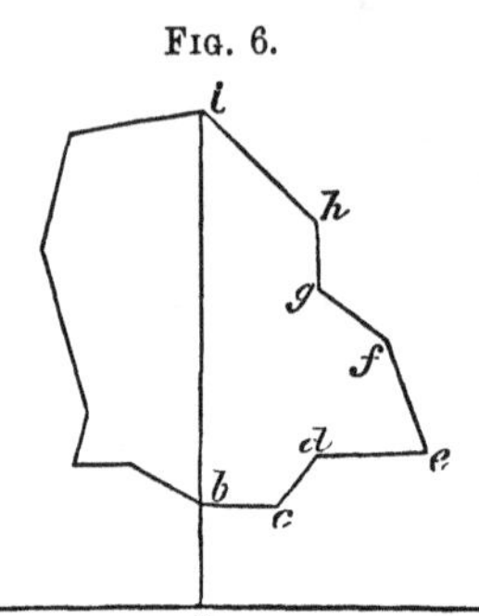

CONVENIENT RULES FOR CORRECTING THE COURSE OF RANDOM LINES, WHEN THE CORRECTION DOES NOT EXCEED 200 LINKS TO EACH MILE.

RULE FOR HALF A MILE, OR FORTY CHAINS.

From the number of links to be corrected in that distance, subtract one-seventh; the difference will be the number of minutes of a degree required for the correction of the course.

Example.

Number of links to be corrected, $42 - 6 = 36'$ answer.

RULE FOR ONE MILE, OR EIGHTY CHAINS.

From half of the number of links to be corrected in that distance, subtract one-seventh, the difference will be the number of minutes of a degree required for the correction of the course.

Example.

Number of links to be corrected, $70 \div 2 = 35 - 5 = 30'$ answer.

RULE FOR THREE MILES.

Divide the whole number of links to be corrected by seven; the quotient will be the number of minutes of a degree required for the correction of the course.

Example.

Number of links to be corrected, $297 \div 7 = 42\frac{3}{7}'$ answer.

RULE FOR SIX MILES.

Divide one-half of the number of links to be corrected by seven; the quotient will be the number of minutes required for the correction of the course.

Example.

Number of links to be corrected, $370 \div 2 = 185 \div 7 = 26\frac{3}{7}'$ answer.*

The distances given for corrections in the above examples, are those for which corrections are generally made in the survey of the public lands, and the calculation for the course of the corrected line can generally be mentally made by the surveyor, while he is occupied in adjusting his instrument.

---

* The above rules are close approximations.

TABLE VI.

Showing the Difference of Latitude and Departure in running 80 chains, at any course from 1 to 60 minutes.

Minutes.	Links.	Minutes.	Links.	Minutes.	Links.
1	$2\frac{1}{3}$	21	49	41	$95\frac{2}{3}$
2	$4\frac{2}{3}$	22	$51\frac{1}{3}$	42	98
3	7	23	$53\frac{2}{3}$	43	$100\frac{1}{3}$
4	$9\frac{1}{3}$	24	56	44	$102\frac{2}{3}$
5	$11\frac{2}{3}$	25	$58\frac{1}{3}$	45	105
6	14	26	$60\frac{2}{3}$	46	$107\frac{1}{3}$
7	$16\frac{1}{3}$	27	63	47	$109\frac{2}{3}$
8	$18\frac{2}{3}$	28	$65\frac{1}{3}$	48	112
9	21	29	$67\frac{2}{3}$	49	$114\frac{1}{3}$
10	$23\frac{1}{3}$	30	70	50	$116\frac{2}{3}$
11	$25\frac{2}{3}$	31	$72\frac{1}{3}$	51	119
12	28	32	$74\frac{2}{3}$	52	$121\frac{1}{3}$
13	$30\frac{1}{3}$	33	77	53	$123\frac{2}{3}$
14	$32\frac{2}{3}$	34	$79\frac{1}{3}$	54	126
15	35	35	$81\frac{2}{3}$	55	$128\frac{1}{3}$
16	$37\frac{1}{3}$	36	84	56	$130\frac{2}{3}$
17	$39\frac{2}{3}$	37	$86\frac{1}{3}$	57	133
18	42	38	$88\frac{2}{3}$	58	$135\frac{1}{3}$
19	$44\frac{1}{3}$	39	91	59	$137\frac{2}{3}$
20	$46\frac{2}{3}$	40	$93\frac{1}{3}$	60	140

VARIATION OF THE NEEDLE.

1. The angle which the magnetic meridian makes with the true meridian, at any place on the surface of the earth, is called the *variation of the needle* at that place, and is east or west, according as the north end of the needle lies on the east or west side of the true meridian.

2. The variation is different at different places, and even at the same place it does not remain constant for any length of time. The variation is ascertained by comparing the magnetic with the true meridian.

3. If we suppose a line to be traced through those points on the surface of the earth, where the needle points directly north, such a line is called the *line of no variation.* At all places lying on the east of this line, the variation

of the needle is west; at all places lying on the west of it, the variation is east.

4. The public is much indebted to Professor Loomis for the valuable results of many observations and much scientific research on the dip and variation of the needle, contained in the 39th and 42d volumes of Silliman's Journal.

The variation at each place was ascertained for the year 1840; and by a comparison of previous observations and the application of known formulas, the annual motion, or change in variation, at each place, was also ascertained, and both are contained in the tables which follow.

5. If the annual motion was correctly found, and continues uniform, the variation at any subsequent period can be ascertained by simply multiplying the annual motion by the number of years, and adding the product, in the algebraic sense, to the variation in 1840. It will be observed that all variations west are designated by the plus sign; and all variations east, by the minus sign. The annual motions being all west, have all the plus sign.

6. Our first object will be to mark the line, as it was in 1840, of *no variation*. For this purpose we shall make a table of places lying near this line.

Places near the Line of no Variation.

Place.	Latitude.	Longitude.	Variation.	An. Motion.
A Point	40° 53′	80° 13′	0° 00′	+4′·4
Cleveland, Ohio	41 31	81 45	−0 19	4·4
Detroit, Mich.	42 24	82 58	−1 56	4
Mackinaw	45 51	84 41	−2 08	3·9
Marietta, Ohio	39 30	81 28	−1 24	4·3
Charlottesville, Va.	39 02	78 30	+0 19	3·7
Charleston, S. C.	32 42	80 04	−2 44	1·3

At the point whose latitude is 40° 53′, longitude 80° 13′, the variation of the needle was nothing in the year 1840, and the direction of the line of no variation, traced north, was N. 24° 35′ west. The line of no variation, prolonged, passed a little to the east at Cleveland, in Ohio—the variation there being 19′ east. Detroit lay still further to the west of this line, the variation there being 1° 56′ east; and Mackinaw still further to the west, as the variation at that place was 2° 08′ east.

The course of the line of no variation, prolonged southerly, was S. 24° 35′ E. Marietta, Ohio, was west of this line—the variation there being 1° 24′ east. Charlottesville, in Virginia, was a little to the east of it—the variation there being 19′ west; while Charleston, in South Carolina, was on the west — the variation there being 2° 44′ east.

From these results, it will be easy to see about where the line of no variation is traced in our own country.

7. We shall give two additional tables:

Places where the Variation was West.

Places.	Latitude.	Longitude.	Variation.	An. Motion.
Angle of Maine	48° 00′	67° 37′	+ 19° 30′	+ 8′·8
Waterville, Me	44 27	69 32	12 36	5·7
Montreal	45 31	73 35	10 18	5·7
Keesville, N. Y	44 28	73 32	8 51	5·3
Burlington, Vt	44 27	73 10	9 27	5·3
Hanover, N. H	43 42	72 14	9 20	5·2
Cambridge, Mass	42 22	71 08	9 12	5
Hartford, Ct	41 46	72 41	6 58	5
Newport, R. I	41 28	71 21	7 45	5
Geneva, N. Y	42 52	77 03	4 18	4·1
West Point	41 25	74 00	6 52	4
New York City	40 43	71 01	5 34	3·6
Philadelphia	39 57	75 11	4 08	3·2
Buffalo, N. Y	42 52	79 06	1 37	4·1

Places where the Variation was East.

Places.	Latitude.	Longitude.	Variation.	An. Motion.
Mouth of Columbia river...	46° 12′	123° 30′	—21° 40′	Unknown.
Jacksonville, Ill..............	39 43	90 20	8 28	+ 2′·5
St. Louis, Mo..................	38 37	90 17	8 37	2·3
Nashville, Tenn...............	36 10	86 52	6 42	2
Louisiana, at..................	29 40	94 00	8 41	1·4
Mobile, Ala.....................	30 42	88 16	7 05	1·4
Tuscaloosa, Ala...............	33 12	87 43	7 26	1·6
Columbus, Geo................	32 28	85 11	5 28	2
Milledgeville, Geo...........	33 07	83 24	5 07	2·4
Savannah, Geo................	32 05	81 12	4 13	2·7
Tallahassee, Fla..............	30 26	84 27	5 03	1·8
Pensacola, Fla................	30 24	87 23	5 53	1·4
Logansport, Ind..............	40 45	86 22	5 24	2·7
Cincinnati, Ohio..............	39 06	84 27	4 46	3·1

METHODS OF ASCERTAINING THE VARIATION.

8. The best practical method of determining the true meridian of a place, is by observing the north star. If this star were precisely at the point in which the axis of the earth, prolonged, pierces the heavens, then, the intersection of the vertical plane passing through it and the place, with the surface of the earth, would be the true meridian. But the star being at a distance from the pole, equal to 1° 30′ nearly, it performs a revolution about the pole in a circle, the polar distance of which is 1° 30′: the time of revolution is 23 h. and 56 min.

To the eye of an observer, this star is continually in motion, and is due north but twice in 23 h. 56 min.; and is then said to be on the meridian. Now, when it departs from the meridian, it apparently moves east or west, for 5 h. and 59 m., and then returns to the meridian again. When at its greatest distance from the meridian, east or west, it is said to be at its greatest *eastern* or *western* elongation.

The following tables show the times of its greatest eastern and western elongations.

Eastern Elongations.

Days.	April.	May.	June.	July.	August.	Sept.
	H. M.	H. M.	H. M.	H. M.	H. M.	H. M.
1........................	18 18	16 26	14 24	12 20	10 16	8 20
7........................	17 56	16 03	14 00	11 55	9 53	7 58
13........................	17 34	15 40	13 35	11 31	9 30	7 36
19........................	17 12	15 17	13 10	11 07	9 08	7 15
25........................	16 49	14 53	12 45	10 43	8 45	6 53

Western Elongations.

Days.	October.	Nov.	Dec.	Jan.	Feb.	March.
	H. M.	H. M.	H. M.	H. M.	H. M.	H. M.
1........................	18 18	16 22	14 19	12 02	9 50	8 01
7........................	17 56	15 59	13 53	11 36	9 26	7 38
13........................	17 34	15 35	13 27	11 10	9 02	7 16
19........................	17 12	15 10	13 00	10 44	8 39	6 54
25........................	16 49	14 45	12 34	10 18	8 16	6 33

The eastern elongations are put down from the first of April to the first of October; and the western, from the first of October to the first of April; the time is computed from 12 at noon. The western elongations in the first case, and the eastern in the second, occurring in the day-time, cannot be used. Some of those put down are also invisible, occurring in the evening, before it is dark, or after daylight in the morning. In such case, if it be necessary to determine the meridian at that particular season of the year, let 5 h. and 59 m. be added to, or subtracted from, the time of greatest eastern or western elongation, and the observation be made at night, when the star is on the meridian.

9. The following table exhibits the angle which the meridian plane makes with the vertical plane passing through the pole-star, when at its greatest eastern or western elongation: such angle is called the *azimuth.* The

mean angle only is put down, being calculated for the first of July of each year.

Azimuth Table.

Year.	Lat. 32° Azimuth.	Lat. 34° Azimuth.	Lat. 36° Azimuth.	Lat. 38° Azimuth.	Lat. 40° Azimuth	Lat. 42° Azimuth.	Lat. 44° Azimuth.
1851.........	1° 45½′	1° 48′	1° 50½′	1° 53½′	1° 56¾′	2° 00¼′	2° 04¼′
1852.........	1° 45′	1° 47½′	1° 50′	1° 53′	1° 56¼′	1° 59¾′	2° 03¾′
1853.........	1° 44½′	1° 47′	1° 49¾′	1° 52½′	1° 55¾′	1° 59¼′	2° 03¼′
1854.........	1° 44¼′	1° 46½′	1° 49¼′	1° 52′	1° 55¼′	1° 59′	2° 02¾′
1855.........	1° 43¾′	1° 46¼′	1° 48¾′	1° 51¾′	1° 54¾′	1° 58½′	2° 02¼′
1856.........	1° 43¼′	1° 45¾′	1° 48¼′	1° 51¼′	1° 54½′	1° 58′	2° 01¾′
1857.........	1° 43′	1° 45¼′	1° 48′	1° 50¾′	1° 54′	1° 57½′	2° 01¼′
1858.........	1° 42½′	1° 44¾′	1° 47½′	1° 50¼′	1° 53½′	1° 57′	2° 00¾′
1859.........	1° 42′	1° 44½′	1° 47′	1° 49¾′	1° 53′	1° 56½′	2° 00¼′
1860.........	1° 41¾′	1° 44′	1° 46½′	1° 49½′	1° 52½′	1° 56′	2° 00′
1861.........	1° 41¼′	1° 43¾′	1° 46¼′	1° 49′	1° 52¼′	1° 55¾′	1° 59½′

The use of the above tables, in finding the true meridian, will soon appear.

TO FIND THE TRUE MERIDIAN WITH THE THEODOLITE.

10. Take a board, of about one foot square, paste white paper upon it, and perforate it through the center; the diameter of the hole being somewhat larger than the diameter of the telescope of the theodolite. Let this board be so fixed to a vertical staff as to slide up and down freely; and let a small piece of board, about three

inches square, be nailed to the lower edge of it, for the purpose of holding a candle.

About twenty-five minutes before the time of the greatest eastern or western elongation of the pole-star, as shown by the tables of elongations, let the theodolite be placed at a convenient point and leveled. Let the board be placed about one foot in front of the theodolite, a lamp or candle placed on the shelf at its lower edge; and let the board be slipped up or down, until the pole-star can be seen through the hole. The light reflected from the paper will show the cross hairs in the telescope of the theodolite.

Then, let the vertical spider's line be brought exactly upon the pole-star, and, if it is an eastern elongation that is to be observed, and the star has not yet reached the most easterly point, it will move from the line toward the east, and the reverse when the elongation is west.

At the time the star attains its greatest elongation, it will appear to coincide with the vertical spider's line for some time, and then leave it, in the direction contrary to its former motion.

As the star moves toward the point of greatest elongation, the telescope must be continually directed to it, by means of the tangent-screw of the vernier plate; and when the star has attained its greatest elongation, great care should be taken that the instrument be not afterward moved.

Now, if it be not convenient to leave the instrument in its place until daylight, let a staff, with a candle or small lamp upon its upper extremity, be arranged at thirty or forty yards from the theodolite, and in the same vertical plane with the axis of the telescope. This is easily effected, by revolving the vertical limb about its horizontal axis without moving the vernier plate, and aligning the staff to coincide with the vertical hair. Then mark the point directly under the theodolite; the line passing through this point and the staff makes an angle with the true meridian equal to the azimuth of the pole-star.

From the table of azimuths, take the azimuth corresponding to the year and nearest latitude. If the observed elongation was east, the true meridian lies on the west of the line which has been found, and makes with it an angle equal to the azimuth. If the elongation was west, the true meridian lies on the east of the line; and, in either case, laying off the azimuth angle with the theodolite, gives the true meridian.

TO FIND THE TRUE MERIDIAN WITH THE COMPASS.

11. 1st. Drive two posts firmly into the ground, in a line nearly east and west; the uppermost ends, after the posts are driven, being about three feet above the surface, and the posts about four feet apart; then lay a plank, or piece of timber three or four inches in width, and smooth on the upper side, upon the posts, and let it be pinned or nailed, to hold it firmly.

2d. Prepare a piece of board four or five inches square, and smooth on the under side. Let one of the compass-sights be placed at right angles to the upper surface of the board, and let a nail be driven through the board, so that it can be tacked to the timber resting on the posts.

3d. At about twelve feet from the stakes, and in the direction of the pole-star, let a plumb be suspended from the top of an inclined stake or pole. The top of the pole should be of such a height that the pole-star will appear about six inches below it; and the plumb should be swung in a vessel of water to prevent it from vibrating.

This being done, about twenty minutes before the time of elongation, place the board, to which the compass-sight is fastened, on the horizontal plank, and slide it east or west, until the aperture of the compass-sight, the plumb-line, and the star are brought into the same range. Then if the star depart from the plumb-line, move the compass-sight east or west along the timber, as the case may be, until the star shall attain its greatest elongation, when it will continue behind the plumb-line for several minutes, and will then recede from it in the direction contrary to its motion before it became stationary. Let the compass-sight be now fastened to the horizontal plank. During this observation it will be necessary to have the plumb-line lighted; this may be done by an assistant holding a candle near it.

Let now a staff, with a candle or lamp upon it, be placed at a distance of thirty or forty yards from the plumb-line, and in the same direction with it and the compass-sight. The line so determined makes, with the true meridian, an angle equal to the azimuth of the pole-star; and from this line the variation of the needle is readily determined, even without tracing the true meridian on the ground.

Place the compass upon this line, turn the sights in the direction of it, and note the angle shown by the needle. Now, if the elongation, at the time of observation, was west, and the north end of the needle is on the west side of the line, the azimuth, plus the angle shown by the needle, is the true variation. But should the north end of the needle be found on the east side of the line, the elongation being west, the difference between the azimuth and the angle would show the variation, and the reverse when the elongation is east.

1. Elongation west, azimuth	2° 04′
North end of the needle on the west, angle . . .	4° 06′
Variation	6° 10′ west.
2. Elongation west, azimuth	1° 59′
North end of the needle on the east, angle . . .	4° 50′
Variation	2° 51′ east.

3. Elongation east, azimuth	2° 05′
North end of the needle on the west, angle . . .	8° 30′
Variation	6° 25′ west.
4. Elongation east, azimuth	1° 57′
North end of the needle on the east, angle . . .	8° 40′
Variation	10° 37′ east.

Remark I. The variation at West Point, in September, 1835, was 6° 32′ west.

Remark II. The variation of the needle should always be noted on every survey made with the compass, and then if the land be surveyed at a future time, the old lines can always be re-run.

12. It has been found by observation, that heat and cold sensibly affect the magnetic needle, and that the same needle will, at the same place, indicate different lines at different hours of the day.

If the magnetic meridian be observed early in the morning, and again at different hours of the day, it will be found that the needle will continue to recede from the meridian as the day advances, until about the time of the highest temperature, when it will begin to return, and at evening will make the same line as in the morning. This change is called the *diurnal variation*, and varies, during the summer season, from one-fourth to one-fifth of a degree.

13. A very near approximation to a true meridian, and consequently to the variation, may be had, by remembering that the pole-star very nearly reaches the true meridian, when it is in the same vertical plane with the star Alioth in the tail of the Great Bear, which lies nearest the four stars forming the quadrilateral.

The vertical position can be ascertained by means of a plumb-line. To see the spider's lines in the field of the telescope at the same time with the star, a faint light should be placed near the object-glass. When the plumb-line, the star Alioth, and the north star fall on the vertical spider's line, the horizontal limb is firmly clamped, and the telescope brought down to the horizon; a light, seen through a small aperture in a board, and held at some distance by an assistant, is then moved according to signals, until it is covered by the intersection of the spider's lines. A picket driven into the ground, under the light, serves to mark the meridian line for reference by day, when the angle formed by it and the magnetic meridian may be measured.

Fig. 7.

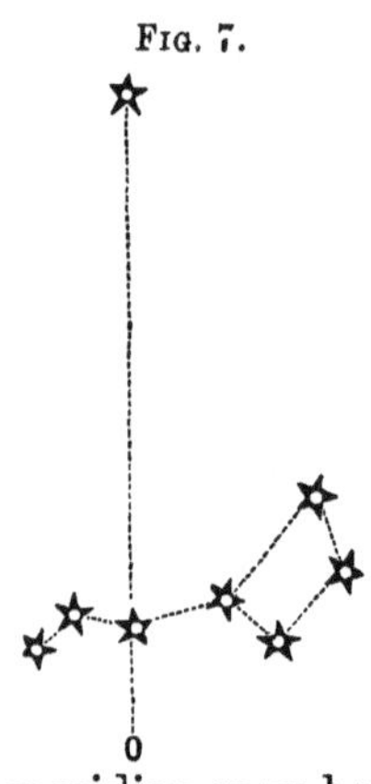

WHEN DESIRED BY SETTLERS, SURVEYS MAY BE MADE BY THE SURVEYOR-GENERAL AT THEIR EXPENSE IN CERTAIN CASES.

It sometimes happens that for the purpose of perfecting their title at an earlier day, the settlers on the public lands prefer to pay for the survey of the township in which their claims are located, in order that it may be surveyed before the government is ready to make the survey in the regular progress of the public work. Congress has provided that this may be done when such townships are contiguous to some established lines of the public surveys from which the regular continuity of the surveys can be secured.

By section 10 of an act entitled "An act to reduce the expenses of the survey and sale of the public lands in the United States," approved May 30th, 1862, it is provided: "That when the settlers in any township or townships, not mineral or reserved by government, shall desire a survey made of the same under the authority of the Surveyor-General of the United States, and shall file an application therefor in writing, and deposit in a proper United States depository, to the credit of the United States, a sum sufficient to pay for such survey, together with all expenses incident thereto, without cost or claim for indemnity on the United States, it shall and may be lawful for said Surveyor-General, under such instructions as may be given him by the Commissioner of the General Land Office, and in accordance with existing laws and instructions, to survey such township or townships, and make return therof to the general and proper local land office: *Provided*, The townships so proposed to be surveyed are within the range of the regular progress of the public surveys embraced by existing standard lines or bases for the township and subdivisional surveys."—(Sec. 10, p. 410, vol. xii. U. S. Laws.)

Applications for surveys under this law must be made to the Surveyor-General in writing, upon the receipt of which he will furnish the applicant with an estimate of how much the desired survey will cost. On receiving a certificate of deposit of a United States depositary, showing that the required sum has been deposited with him in a proper manner to pay for the work, the Surveyor-General will contract with a competent United States deputy surveyor, and have the survey made and returned in the same manner as other public surveys are.

The Surveyors-General are especially enjoined in all cases to state explicitly in their letters furnishing estimates to applicants, that the payment of the amount required for the survey will not give the depositor any priority of claim or right to purchase the land, or in any manner affect the claim or claims of any party or parties thereto; and that, when surveyed, it will be subject to the same laws and regulations in relation to the disposition thereof that other public lands are.

The money should be deposited to the credit of the Treasurer of the United States on account of the proper appropriations. A separate estimate is required and a separate deposit must be made for *office work* and *field work;* one to be placed to the credit of the appropriation "for compensation of the Surveyor-General and the clerks in his office," and the other to the credit of the appropriation "for continuing the public surveys." The depositary will issue certificates in triplicate, one of which will be transmitted to the General Land Office with the contract and bond of the deputy surveyor.

The account will be adjusted and paid in the same manner as other surveying accounts. Should the amount deposited exceed the cost of survey and all expenses incident thereto, including office work, an account setting forth the fact of such excess may be rendered by the depositor, certified by the Surveyor-General, and transmitted to the General Land Office, with the final surveying returns, to be recorded for payment.

Where a township is surveyed under the provisions of the aforesaid act, the survey must include all the *surveyable* public land in such township.

CERTAIN SMALL ISLANDS MAY BE SURVEYED AT THE COST OF APPLICANTS.

In the early extension of the public surveys it frequently happened that small islands were omitted when the adjacent lands were surveyed. In the course of time as the country is settled these islands become more valuable, and numerous letters are received at the General Land Office from parties desiring to purchase, asking how they shall proceed to acquire a title to them.

The first step to be taken is to have such islands surveyed by an authorized government surveyor. When the islands are located in a surveying district where there is a Surveyor-General, the application for a survey should be made directly to that officer; should the applicant address the General Land Office he will invariably be referred to the local office in such cases.

The Commissioner of the General Land Office is by law *ex officio* Surveyor-General for those older land states, where the public lands have been surveyed and the Surveyor-General's office has been closed; applications for the survey of islands in these states should therefore be made to said Commissioner.

These islands are usually of too little value to justify the government in incurring the expense of survey; but where a party desires the survey made and is willing to pay the cost thereof in advance, upon the conditions set forth in these instructions, it may be done under the provisions of the 10th section of the act of May 30th, 1862.

The party desiring the survey to be made must file a written application with the Surveyor-General, giving an intelligible description of the locality of the island, its distance from the main shore, the width of the narrowest channel between it and the main land, with an estimate of its area.

Upon receiving such application, made in the manner indicated, the Surveyor-General will examine the records and dates in his office, and if it appears that the island is public land and has not already been surveyed, he will furnish the applicant an estimate of the cost of surveying it, as directed for carrying into effect the 10th section of the act of May 30th, 1862, observing particularly that two separate deposits are to be made—one on account of the appropriation for *field work*, and one on account of the appropriation for *office work*—a separate certificate for each to be transmitted to the General Land Office with the contract and bond of the deputy surveyor.

It will be understood that these instructions relate only to isolated islands, or islands that were omitted when the public surveys were extended over the adjacent lands, and do not apply to islands falling within the regular course of current surveys, which must be included in the contracts for surveying the public lands.

As a general rule, a body of land separated from the main land by a *perpetual* natural channel may be regarded as an island for the purposes contemplated in these instructions.

It will be understood, also, that the paying for the survey of these islands in advance does not give any priority of claim to the party depositing the money, nor in any manner affect the rights of parties in said lands. The survey is a necessary first step to get the land in a condition to be disposed of, and when it is surveyed it becomes subject to the same laws and regulations that other public lands are. It is the duty of Surveyors-General so to advise applicants before receiving their money.

FIELD NOTES

OF THE

EXTERIOR AND SUBDIVISION LINES

OF TOWNSHIP NO. ——, RANGE NO. ——, OF THE ——TH PRINCIPAL MERIDIAN.

MINNESOTA

SURVEYED BY —— ——, DEPUTY SURVEYOR, UNDER HIS CONTRACT, DATED ——, 18——.

SURVEY COMMENCED ——.

SURVEY COMPLETED ——

DIAGRAM.

INDEX REFERRING THE LINES TO THE PAGES OF THE FIELD NOTES.

Township No. 25 N. Range No. 2 W. Willamette Meridian.

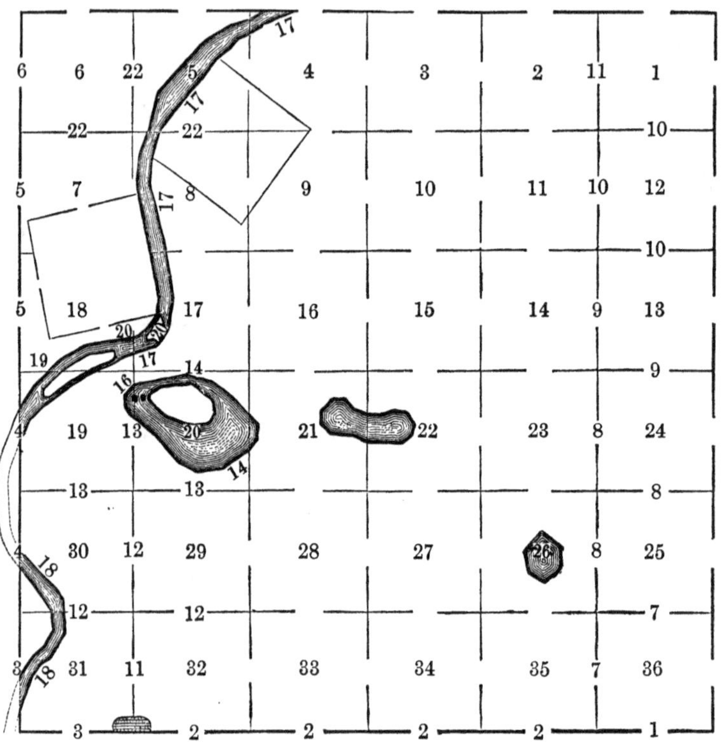

FIELD NOTES.

South Boundary, T. 25 N. R. 2 W. Willamette Meridian.

CHAINS. Begin at the post, the established corner to Townships 24 and 25 North, in Ranges 2 and 3 West. The witness trees all standing, and agree with the description furnished me by the office, viz.:

A Black Oak, 20 in. dia. N. 37 E. 27 links,
A Bur Oak, 24 in. dia. N. 43 W. 35 links,
A Maple, 18 in. dia. S. 27 W. 39 links,
A White Oak, 15 in. dia. S. 47 E. 41 links.

East, on a random line on the South Boundaries of sections 31, 32, 33, 34, 35, and 36.

Variation by Burt's improved solar compass, 18° 41′ E,

I set temporary half mile and mile posts at every 40 and 80 chains, and at 5 miles, 74 chains, 53 links, to a point 2 chains and 20 links north of the corner to Townships 24 and 25 North, Ranges 1 and 2 W.

(Therefore the correction will be 5 chains, 47 links *West*, and 37 links *South* per mile),

I find the corner post standing and the witness trees to agree with the description furnished me by the surveyor-general's office, viz.:

A Bur Oak, 17 in. dia. bears N. 44 E. 31 links,
A White Oak, 16 in. dia. N. 26 W. 21 links,
A Lynn, 20 in. dia. S. 42 W. 15 links,
A Black Oak, 24 in. dia. S. 27 E. 14 links.

From the corner to Townships 24 and 25 N. Ranges 1 and 2 West, I run North 89° 44′ west, on a *true* line along the South Boundary of section 36,

Variation 18° 41′ East,

40·00 Set a post for quarter section corner, from which

A Beech, 24 in. dia. bears N. 11 E. 38 links dist.
A Beech, 9 in. dia. bears S. 9 E. 17 links dist.

62·50 A Brook, 6 links wide runs North,

80·00 Set a post for corner to sections 35 and 36, 1 and 2, from which

A Beech, 9 in. dia. bears N. 22 E. 16 links dist.
A Beech, 8 in. dia. bears N. 19 W. 14 links dist
A White Oak, 10 in. dia. bears S. 52 W. 7 links dist.
A Black Oak, 14 in. dia. bears S. 46 E. 8 links dist.

Land level, good soil, fit for cultivation,

Timber, Beech; various kinds of Oak, Ash, and Hickory.

[2]

South Boundary, T. 25 N. R. 2 W. Willamette Meridian.

CHAINS.	North 89° 44′ west, on a *true* line along the South Boundary of section 35,
	Variation 18° 41′,
40·00	Set a post for quarter section corner, from which
	A Beech, 8 in. dia. bears N. 20 E. 8 links dist.
	No other tree convenient; made trench around post,
65·00	Begin to ascend a moderate hill; bears N. and S.
80·00	Set a post with trench, for corner of sections 34 and 35, 2 and 3, from which
	A Beech, 10 in. dia. bears N. 56 W. 9 links dist.
	A Beech, 10 in. dia. bears S. 51 E. 13 links dist.
	No other trees convenient to mark,
	Land level, or gently rolling, and good for farming,
	Timber, Beech, Oak, Ash, and Hickory; some Walnut and Poplar.
	North 89° 44′ west, on a *true* line along the South Boundary of section 34,
	Variation 18° 41′,
40·00	Set a quarter section post with trench, from which
	A Black Oak, 10 in. dia. bears N. 2 E. 635 links dist.
	No other tree convenient to mark,
80·00	To point for corner of sections 33, 34, 3 and 4,
	Drove charred stakes, raised mounds with trenches, as per instructions, from which
	A Bur Oak, 16 in. dia. bears N. 31 E. 344 links, and
	A Hickory, 12 in. dia. bears S. 43 W. 231 links,
	No other trees convenient to mark,
	Land level, rich and good for farming,
	Timber, some scattering Oak and Walnut.
	North 89° 44′ west, on a *true* line along the South Boundary of section 33,
	Variation 18° 41′,
37·51	A Black Oak, 24 in. dia.
40·00	Set a post for quarter section corner, from which
	A Black Oak, 18 in. dia. bears N. 25 E. 32 links dist.
	A White Oak, 15 in. dia. bears N. 43 W. 22 links dist.
62·00	To foot of steep hill, bears N. E. and S. W.
80·00	Set a post for corner to sections 32, 33, 4 and 5, from which
	A White Oak, 15 in. dia. bears N. 23 E. 27 links dist.
	A Black Oak, 20 in. dia. bears N. 82 W. 75 links dist.
	A Bur Oak, 20 in. dia. bears S. 37 W. 92 links dist.
	A White Oak, 24 in. dia. bears S. 26 E. 42 links dist.
	Land gently rolling; good rich land for farming,
	Timber, Black and White Oak, Hickory, and Ash.
	North 89° 44′ west, on a *true* line along the South Boundary of section 32,
	Variation 18° 41′,
37·50	A creek 20 links wide, runs North,
40·00	Set a granite stone 14 in. long, 10 in. wide, and 4 in. thick, for quarter section corner from which

[3]

South Boundary, T. 25 N. R. 2 W. Willamette Meridian.

CHAINS.	
	A Maple, 20 in. dia. bears N. 41 E. 25 links distant,
	A Birch, 24 in. dia. bears N. 35 W. 22 links distant,
76·00	To S. E. edge of swamp,
	As it is impossible to establish *permanently* the corner to sections 31, 32, 5 and 6 in the swamp, I therefore at this point, 400 chains East of the true point for said section cor. raise a witness mound with trench, as per instructions, from which
	A Black Oak, 20 in. dia. bears N. 51 E. 115 links,
80·00	A point in deep swamp for corner to sections 31, 32, 5 and 6,
	Land, rich bottom; *west* of creek part wet; *east* of creek good for farming,
	Timber, good; Oak, Hickory, and Walnut.
	North 89° 44′ West, on a *true* line along the South Boundary of section 31,
	Variation 18° 41′,
11·00	Leave swamp and rise bluff 30 feet high, bears N. and S.
40·00	Set post for quarter section corner, from which
	A Sugar-Tree, 27 in. dia. bears S. 81 W. 42 links dist.
	A Beech, 24 in. dia. bears S. 71 E. 24,
54·00	Foot of rocky bluff 30 feet high, bears N. E. and S. W.
57·50	A spring branch comes out at the foot of the bluff 5 links wide; runs N. W. into swamp,
61·00	Enter swamp; bears N. and S.
70·00	Leave swamp; bears N. S. The swamp contains about 15 acres, the greater part in section 31,
74·73	The corner to Townships 24 and 25 N. Ranges 2 and 3 W.
	Land, except the swamp, rolling, good, rich soil,
	Timber, Sugar-Tree, Beech, and Maple.
	January 25th, 1854.
	From the corner to Townships 24 and 25 N. Ranges 2 and 3 West, I run
	North, on the Range line between sections 31 and 36,
	Variation 18° 56′ East,
8·56	Set a post on the left bank of Chickeeles River, for corner to fractional sections 31 and 36, from which
	A Hackberry, 11 in. dia. bears N. 50 E. 11 links dist.
	A Sycamore, 60 in. dia. bears S. 15 W. 24 links dist.
	I now cause a flag to be set on the *right* bank of the river, and in the line between sections 31 and 36. I now cross the river, and from a point on the right bank thereof, *West* of the corner just established on the left bank, I run *North*, on an offset line, 25 chains and 94 links, to a point 8 chains and 56 links *West* of the flag. I now set a post in the place of the flag, for corner to fractional sections 31 and 36, from which
	A Beech, 10 in. dia. bears N. 2 E. 12 links dist.
	A Black Oak, 12 in. dia. bears N. 80 W. 16 links dist.
34·50	The corner above described,
40·00	Set a post for quarter section corner, from which
	A Bur Oak, 20 in. dia. bears N. 37 E. 26 links dist.

[4]

Between Ranges 2 and 3 W. T. 25 N. Willamette Meridian.

CHAINS.	
	A Black Oak, 24 in. dia. bears S. 75 W. 21 links dist.
43·41	A Black Walnut, 30 in dia.
80·00	Set a post for corner to sections 30, 31, 25 and 36, from which
	A Beech, 14. in. dia. bears N. 20 E. 14 links dist.
	A Hickory, 9 in. dia. bears N. 25 W. 12 links dist.
	A Beech, 16 in. dia. bears S. 40 W. 16 links dist.
	A White Oak, 10 in. dia. bears S. 44 E. 20 links dist.
	Land level; rich bottom; not subject to inundation.
	Timber, White and Black Oak, Beech, Hickory; and Ash.
	North, between sections 25 and 30,
	Variation 18° 50′ East,
27·73	Set a post for corner to fractional sections 25 and 30 on the right bank of Chickeeles River, a navigable stream, which here runs S. E. from which
	A Willow, 6 in. dia. bears S. 37 W. 55 links dist.
	A Maple, 20 in. dia. bears S. 30 E. 11 links dist.
	I now cause a flag to be set on the left bank of the river, and in the line between sections 25 and 30. From the above corner I run West 333 chains to a point from which the flag bears N. 16° 30′ E. which gives for the distance across the river on the line 11·27 chains, to which add 27·73, makes
39·00	To the flag on the bank, I here set a post for corner to fractional sections 25 and 30, from which
	A Hickory, 8 in. dia. bears N. 44 E. 17 links dist.
	A White Oak, 8 in. dia. bears N. 15 W. 8 links dist.
40·00	Set a post for quarter section corner, from which
	A Hickory, 9 in. dia. bears N. 16 E. 16 links dist.
	A Buckeye, 10 in. dia. bears S. 16 E. 18 links dist.
43·71	A Hickory, 24 in. dia.
80·00	Set a post for corner to sections 19, 30, 24, 25, from which
	An Elm, 6 in. dia. bears N. 82 E. 25 links dist.
	A Sugar-Tree, 14 in. dia. bears N. 49 W. 4 links dist.
	An Elm, 9 in. dia. bears S. 42 W. 30 links dist.
	A Sugar-Tree, 10 in dia. bears S. 55 E. 45 links dist.
	Land good; rich bottom, 1st rate,
	Timber, Hickory, Elm, Buckeye, Sugar-Tree, and Ash.
	North, between sections 19 and 24,
	Variation 18° 50′ East,
32·50	A Hickory, 20 in. dia. on the left bank of Chickeeles River, mark it for corner to fractional sections 19 and 24, from which
	A Hackberry, 20 in. dia. bears S. 13 W. 27 links dist.
	A Black Oak, 24 in. dia. bears S. 27 E. 31 links dist.
	I now cause a flag to be set on the right bank of the river, and in the line between sections 19 and 24, and from the corner run a *base* East 5·90 chains to a point from which the flag bears N. 17 W. continue the base East to a point 9·00 chains *East* of the corner on the river bank, from which the flag bears N. 25° 15′ W. which gives by calculation as the mean result of the two observations for the distance across the

[5]

Between Ranges 2 and 3 W. T. 25 N. Willamette Meridian.

CHAINS.	
	river on the line between sections 19 and 24, 19·30 chains, to which add 32·50 chains, the distance to the river, makes
51·80	To the flag on the right bank of the river; I here set a post for corner to fractional sections 19 and 24, from which
	A Beech, 12 in. dia. bears N. 24 E. 30 links dist.
	A Beech, 14 in. dia. bears S. 55 W. 120 links dist.
	NOTE.—The point for quarter section corner falling in the river, it cannot therefore be established.
55·74	A Black Oak, 30 inches diameter,
80·00	Set a post for corner to sections 18, 19, 13, and 24, from which
	A White Oak, 18 in. dia. bears N. 55 E. 24 links dist.
	A White Oak, 17 in. dia. bears N. 64 W. 18 links dist.
	A Red Oak, 27 in dia. bears S. 26 W. 20 links dist.
	A Red Oak, 15 in. dia. bears S. 29 E. 40 links dist.
	Land good; rich bottom; not subject to inundation.
	Timber, various kinds of Oak, Beech, Hickory, and Ash; undergrowth same, and vines.
	North, between sections 13 and 18,
	Variation 18° 53′ East,
5·00	Leave bottom and enter upland; bears N. E. and S. W.
21·88	A Red Oak, 20 in. dia.
38·60	A White Oak, 24 in. dia.
40·00	Set a post for quarter section corner, from which
	A White Oak, 22 in. dia. bears N. 27 W. 27 links dist.
	A White Oak, 23 in. dia. bears S. 28 E. 92 links dist.
46·50	A road from Williamsburg bears East and West,
68·37	A Black Walnut, 21 in. dia.
80·00	Set a post for corner to sections 7, 18, 12, and 13, from which
	A White Oak, 12 in. dia. bears N. 55 E. 68 links dist.
	A Black Oak, 8 in. dia. bears N. 53 W. 40 links dist.
	A Black Oak, 16 in. dia. bears S. 40 W. 55 links dist.
	A Red Oak, 10 in. dia. bears S. 44 E. 50 links dist.
	Land rolling, and next the bottom broken; soil 2d rate,
	Timber good; various kinds of Oak and Hickory,
	North, between sections 7 and 12,
	Variation 18° 53′ East,
15·18	A White Oak, 15 in. dia.
30·26	A White Oak, 21 in. dia.
40·00	Set a post for quarter section corner, from which
	A White Oak, 12 in. dia. bears S. 13 W. 60 links dist.
	A White Oak, 15 in. dia. bears S. 35 E. 55 links dist.
68·37	A Black Walnut, 21 in. dia.
80·00	Set a post for corner to sections 6, 7, 1, and 12, from which
	A White Oak, 17 in. dia. bears N. 58 E. 60 links dist.
	A White Oak, 18 in. dia. bears N. 54 W. 51 links dist.
	A White Oak, 18 in. dia. bears S. 51 W. 20 links dist.
	A Hickory, 14 in. dia. bears S. 64 E. 42 links dist.
	Land gently rolling, 2d rate.
	Timber, Oak and Hickory; undergrowth, Oak and Hazel.

[6]

Between Ranges 2 *and* 3 *W. T.* 25 *N. Willamette Meridian.*

CHAINS.	North, between sections 1 and 6, Variation 18° 53′ East,
3·00	Enter *stony* barrens; timber scattering; bears East and West,
25·31	A Blackjack, 12 in. dia.
40·00	Set a quartz stone, 13 in. long, 12 in. wide, and 4 in. thick, for quarter section corner, with trench, as per instructions, from which A Blackjack, 20 in. dia. bears S. 44 E. 95 links dist. No other tree convenient to mark,
45·00	Leave *stony* barrens, bears East and West,
61·11	A Hickory, 10 in. dia. Here leave timber and enter prairie, bearing West and N. E.,
80·00	Set a granite stone, 18 in. long, 12 in. wide, and 6 inches thick, for corner to Townships 25 and 26 North, Ranges 2 and 3 West; raise a stone mound, with trench, as per instructions, Land broken and stony; too poor for cultivation, Timber, scattering and poor; Blackjack and Hickory, January 27th, 1851.

GENERAL DESCRIPTION.

This Township contains a large amount of first-rate land for farming. It is well timbered with the various kinds of Oak, Hickory, Sugar-Tree, Walnut, Beech, and Ash.

Chickeeles River is navigable for small boats in low water, and does not often overflow its banks, which are from ten to fifteen feet high.

The Township will admit of a large settlement, and should therefore be subdivided.

[7]

Field Notes of the subdivision lines and meanders of Chickeeles River in Township 25 *N. R.* 2 *W. Willamette Meridian.*

CHAINS.	
	To determine the proper adjustment of my compass for subdividing this Township, I commence at the corner to Townships 24 and 25 N. R. 1 and 2 W., and run
	North, on a blank line along the East Boundary of section 36, Variation 17° 51′ East,
40·05	To a point 5 links West of the quarter section corner,
80·09	To a point 12 links West of the corner to sections 25 and 36,
	To retrace this line or run parallel thereto, my compass must be adjusted to a variation of 17° 46′ East.
	Subdivision commenced February 1, 1854
	From the corner to sections 1, 2, 35 and 36 on the South Boundary of the Township, I run
	North, between sections 35 and 36, Variation 17° 46′ East,
9·19	A Beech, 30 in. dia.
29·97	A Beech, 30 in. dia.
40·00	Set a post for quarter section corner, from which
	A Beech, 8 in. dia. bears N. 23 W. 45 links dist.
	A Beech, 15 in. dia. bears S. 48 E. 12 links dist.
51·00	A Beech, 18 in. dia.
76·00	A Sugar-Tree, 30 in. dia.
80 00	Set a post for corner to sections 25, 26, 35 and 36, from which
	A Beech, 28 in. dia. bears N. 60 E. 45 links dist.
	A Beech, 24 in. dia. bears N. 62 W. 17 links dist.
	A Poplar, 20 in. dia. bears S. 70 W. 50 links dist.
	A Poplar, 36 in. dia. bears S. 66 E. 34 links dist.
	Land level; 2d rate,
	Timber, Poplar, Beech, Sugar-Tree, and some Oak; undergrowth, same and hazel.
	East, on a *random* line between sections 25 and 36, Variation 17° 46′ East,
9·00	A brook, 20 links wide, runs north,
15·00	To foot of hills bears N. and S.
40·00	Set a post for temporary quarter section corner
55·00	To opposite foot of hill, bears N. and S.
72·00	A brook 15 links wide runs North,
80·00	Intersected East Boundary at post corner to sections 25 and 36, from which corner I run
	West, on a *true* line between sections 25 and 36, Variation 17° 46′ East,
40·00	Set a post on top of hill bears N. and S. from which
	A Hickory, 14 in. dia. bears N. 60 E. 27 links dist.
	A Beech, 15 in. dia. bears S. 74 W. 9 links dist.
80·00	The corner to sections 25, 26, 35 and 36,
	Land, east and west parts level, 1st rate; middle part broken, 3d rate,
	Timber, Beech, Oak, Ash, etc.; undergrowth, same and Spice in the branch bottoms.

[8]

Township 25 *N. Range* 2 *W. Willamette Meridian.*

CHAINS.	North, between sections 25 and 26, Variation 17° 46′ East,
7 00	A Poplar, 40 in dia.
17·20	A brook, 25 links wide, runs N. W.
18·05	A Walnut, 30 in. dia.
23·44	A brook, 25 links wide, runs N. E.
40 00	Set a post for quarter section corner, from which
	A Bur Oak, 36 in. dia. bears N. 42 E. 18 links dist.
	A Beech, 30 in. dia. bears S. 72 W. 9 links dist.
60·15	A Beech, 30 in. dia.
80·00	Set a post for corner to sections 23, 24, 25, and 26, from which
	A White Oak, 14 in. dia. bears N. 50 E. 40 links.
	A Sugar-Tree, 12 in. dia. bears N. 14 W. 31 links dist.
	A White Oak, 13 in. dia. bears S. 38 W. 32 links dist.
	A Sugar-Tree, 12 inch dia. bears S. 42 E. 14 links dist.
	Land level on the line, high ridge of hills through the middle of Section 25 running N. and S.
	Timber, Beech, Walnut, Ash, Sugar-Tree, etc.
	East, on a *random* line between sections 24 and 25, Variation 17° 46′ East,
8·90	A stream, 30 links wide, rapid current, runs N. W.
12·00	To foot of hill, bears south and N. E.
40·00	Set a post for temporary quarter section corner,
48·00	To opposite foot of hill, bears south and N. W.
60·50	A stream, 30 links wide, runs N., soon turns N. W.
73·00	To foot of hill, rises moderately, bears S. and N. W.
80·12	Intersected East Boundary of the Township at the post corner to sections 24 and 25, from which corner I run
	West, on a *true* line between sections 24 and 25, Variation 17° 46′ East,
40·06	Set a post for quarter section corner, from which
	A Beech, 18 in. dia. bears N. 74 W. 26 links dist.
	A Beech, 16 in. dia. bears S. 73 E. 22 links dist.
80·12	The corner to sections 23, 24, 25, 26,
	Land rolling between the branches; good, 2d rate; branch bottoms level, 1st rate,
	Timber, Walnut, Beech, Elm, and Oak; undergrowth, same and Spice.
	North, between sections 23 and 24, Variation 17° 46′ East
6·70	A White Oak, 20 in dia.
9·65	A stream, 25 links wide, runs N. W.
13·50	Same stream, 25 links wide, runs N. E.
16·00	Same stream, 25 links wide, runs N. W.
40·00	Set a post near the South bank of a stream for quarter section corner, from which
	A Cottonwood, 18 in. dia. bears S. 7 W. 7 links dist.
	A White Walnut, 24 in. dia. bears S. 22 E. 4 links dist.
40·35	Elk Creek, 125 links wide, runs N. W. general course West.
	John Jones has a field on the North side of the creek and West

[9]

Township 25 *N. Range* 2 *W. Willamette Meridian.*

CHAINS.	
	of the line; his house is 2 chains South of the road and 2 chains East of the line,
54·00	To the road from Astoria to Williamsburg, bears E. and W.
58·00	Enter wet prairie, bears East and West,
68 00	Leave prairie and enter timber, bearing East and West,
	This prairie extends *East* into section 24 about 30 chains,
72·12	A White Oak, 30 in. dia.
75·00	Leave creek bottom and enter hills bearing East and West,
80·00	Set a post for corner to sections 13, 14, 23, 24, from which
	A White Walnut, 16 in. dia. bears N. 42 E. 15 links dist.
	A White Walnut, 24 in. dia. bears N. 59 W. 27 links dist.
	An Elm, 8 in. dia. bears S. 67 W. 16 links dist.
	A Black Oak, 14 in. dia. bears S. 38 E. 17 links dist.
	Land mostly level; 1st rate soil,
	Timber, Walnut, various kinds of Oak, Buckeye, and Hickory; undergrowth, same and Spice,
	February 1st, 1854.
	East, on a *random* line between sections 13 and 24,
	Variation 17° 46′ East,
40·00	Set a post for temporary quarter section corner,
80·10	Intersected the East Boundary of Township 16 links South of post corner to sections 13 and 24, from which corner I run
	North 89° 53′ west, on a *true* line between sections 13 and 24,
	Variation 17° 46′ East,
40 05	Set a post for quarter section corner, from which
	A Sugar-Tree, 30 in. dia. bears N. 80 W. 22 links dist.
	A White Oak, 16 in. dia. bears S. 53 E. 20 links dist.
80·10	The corner to sections 13, 14, 23, 24,
	Land mostly rolling; good rich soil; 1st rate,
	Timber, Walnut, Sugar-Tree, Oak, Elm, and Buckeye; undergrowth, same and Spice.
	North, between sections 13 and 14,
	Variation 17° 46′ East,
6·17	A White Oak, 30 in. dia.
22 15	A Beech, 30 in. dia.
40·00	Set a post for quarter section corner, from which
	A Beech, 24 in. dia. bears N. 66 W. 6 links dist.
	A Beech, 20 in. dia. bears S. 45 E. 40 links dist.
52·25	A Beech, 24 in. dia.
62·61	A Bur Oak, 30 in. dia.
80·00	Set a post for corner to sections 11, 12, 13, 14, from which
	A Black Oak, 26 in. dia. bears N. 53 E. 10 links dist.
	A Black Oak, 21 in. dia. bears N. 20 W. 35 links dist.
	A Sugar-Tree, 30 in. dia. bears S. 32 W. 25 links dist.
	A White Oak, 20 in. dia. bears S. 24 E. 20 links dist.
	Land gently rolling; good, 2d rate,
	Timber, Beech, Oak, and Ash; undergrowth, same and Hazel.

[10]

Township 25 N. Range 2 W. Willamette Meridian.

CHAINS.	East, on a *random* line between sections 12 and 13,
	Variation 17° 46′ East,
20·50	Foot of hills, and enter broken ridges bearing North and South,
40·00	Set a post for temporary quarter section corner,
80·10	Intersected East Boundary 13 links North of post corner to sections 12 and 13, from which corner I run
	North 89° 54′ west, on a *true* line between sections 12 and 13,
	Variation 17° 46′ East,
40·05	Set a post for quarter section corner, from which
	An Elm, 24 in. dia. bears N. 51 E. 50 links dist.
	A Beech, 18 in. dia. bears S. 51 W. 29 links dist.
80·10	The corner to sections 11, 12, 13, 14,
	Land West 20 chains; gently rolling; good, 2d rate; the balance high, broken ridges,
	Timber, Beech, Black Oak, and White Oak; undergrowth, same and Hazel.
	North, between sections 11 and 12,
	Variation 17° 46′ East,
10·81	An Elm, 15 in dia.
40·00	Set a post for quarter section corner, from which
	A Beech, 30 in. dia. bears N. 33 W. 9 links dist.
	A Beech, 20 in. dia. bears S. 64 W. 20 links dist.
52·25	A Beech, 24 in. dia.
62·61	A Black Oak, 30 in. dia.
75·40	A spring branch, 10 links wide, runs West,
80·00	Set a post for corner to sections 1, 2, 11 and 12, from which
	A Poplar, 32 in. dia. bears N. 41 E. 30 links dist.
	A Poplar, 36 in. dia. bears N. 43 W. 25 links dist.
	A Sugar-Tree, 30 in. dia. bears S. 32 W. 25 links dist.
	A Sugar-Tree, 31 in. dia. bears S. 35 E. 40 links dist
	Land level; good, 2d rate,
	Timber, Sugar-Tree, Poplar, Walnut, and Oak; undergrowth, same and Hazel.
	East, on a *random* line between sections 1 and 12,
	Variation 17° 46′ East,
23·00	Enter high, broken ridges, bearing N. E. and South,
40 00	Set a post for temporary quarter section corner,
42·50	A spring branch, 10 links wide, runs S. W.
63·00	To foot of high mountain; bears North and South,
80·24	Intersected the East Boundary of the Township 13 links North of post corner to sections 1 and 12, from which corner I run
	West, on a *true* line between sections 1 and 12,
	Variation 17° 40′ East,
40·12	Set a post on top of narrow ridge, bearing North and South, for quarter section corner, from which
	A Sugar-Tree, 20 in. dia. bears N. 20 E. 32 links dist.
	A Sugar-Tree, 24 in. dia. bears S. 56 W. 25 links dist.
80·24	The corner to sections 1, 2, 11, 12,
	Land very broken and mountainous,
	Timber, Sugar-Tree, Beech; various kinds of Oak and Hickory,

[11]

Township 25 N. Range 2 W. Willamette Meridian.

CHAINS.	
	☞ On this line, and towards the foot of the mountain, we discovered gold dust; and throughout the line we observed many specimens of what appeared to be rich auriferous quartz.
	North, on a *random* line between sections 1 and 2,
	Variation 17° 46′ East,
40·00	Set a post for temporary quarter section corner,
80·11	Intersected the North Boundary 32 links East of corner to sections 1 and 2, from which corner I run
	South 14′ East, on a *true* line between sections 1 and 2,
	Variation 17° 46′ East,
40·11	Set a post for quarter section corner, from which
	A White Oak, 20 in. dia. bears N. 31 W. 65 links dist.
	A Sugar-Tree, 14 in. dia. bears S. 49 E. 32 links dist.
80·11	The corner to sections 1, 2, 11, 12,
	Land level; good, rich soil.
	Timber, Walnut, Sugar-Tree, Beech, and various kinds of Oak; open woods.
	February 2d, 1854.
	The point for corner to sections 5, 6, 31 and 32 being in a deep swamp, and not having been established, I begin at the witness corner on the S. E. edge of the swamp, 4·00 chains East of said point, and run thence East 250 links (with the line between sections 5 and 32) to a point; thence *North* 7·50 chains to a point; thence *West* 6·50 chains to a point on the North edge of the swamp and in the line between sections 31 and 32, and 7·50 chains *North* of the point for corner to sections 31 and 32, on the South Boundary of the Township. I here set a post for witness point, from which
	A Bur Oak, 16 in. dia. bears N. 31 E. 25 links dist
	An Ash, 12 in. dia. bears N. 25 W. 17 links dist.
	From this witness point I run
	North, between sections 31 and 32, counting the distance from the point for corner to said sections in the swamp,
	Variation 17° 40′ East,
12·98	A Walnut, 22 in. dia.
38·19	An Ash, 35 in. dia.
40·00	Set a post for quarter section corner, from which
	A Beech, 20 in. dia. bears N. 12 W. 45 links dist.
	A Sugar-Tree, 20 in. dia. bears S. 12 E. 13 links dist.
57·74	An Ash, 24 in. dia.
66·19	A White Oak, 36 in. dia.
80·00	Set a post with trench for corner to sections 29, 30, 31, 32, from which
	A Beech, 26 in. dia. bears N. 9 W. 12 links dist.
	A Sugar-Tree, 24 in. dia. bears S. 13 E. 56 links dist.
	And planted N. E. a Butternut, and S. W. 4 Cherry stones.
	Land, South half level, North half rolling; good soil,
	Timber, Oak, Beech, Sugar-Tree, and Walnut; undergrowth, same and Hazel on North part.

[12]

Township 25 *N. Range* 2. *W. Willamette Meridian.*

CHAINS.	
	East, on a *random* line between sections 29 and 32, Variation 17° 40′ East,
40·00	Set a post for temporary quarter section corner,
80·16	Intersected the N. and S. line 10 links N. of post corner to sections 28, 29, 32 and 33, from which corner I run
	North 89° 56′ West, on a *true* line between sections 29 and 32, Variation 17° 40′ E.
40·08	Set a post for quarter section corner, from which A Black Oak, 18 in. dia. bears N. 36 E. 42 links dist. A Bur Oak, 20 in. dia. bears S. 43 W. 47 links dist.
80·16	The corner to sections 29, 30, 31, 32,
	Land gently rolling; good soil; fit for cultivation,
	Timber, Oak, Beech, Hickory, and Walnut; open woods.

	West, on a *true* line between sections 30 and 31, knowing that it will strike the Chickeeles River in less than 80·00 chains, Variation 17° 40′ East,
3·41	A Whike Oak, 15 in. dia.
5·00	Leave upland and enter creek bottom, bearing N. E. and S. W.
8·00	Elk Creek, 200 links wide, gentle current, muddy bottom and banks, runs S. W.
	Ascertain the distance across the creek on the line as follows, viz.:
	Cause the flag to be set on the right bank of the creek, and in the line between sections 30 and 31. From the station on the left bank of creek, at 8·00 chains, I run *South* 245 links to a point from which the flag on the right bank bears N. 45 W. which gives for the distance across the creek, on the line between sections 30 and 31, 2 chains 45 links.
25·17	A Bur Oak, 24 in. dia.
40·00	Set a post for quarter section corner, from which A Buckeye, 24 in. dia. bears N. 15 W. 8 links dist. A White Oak, 30 in. dia. bears S. 65 E. 12 links,
41·90	Set a post on the left bank of Chickeeles River, a navigable stream, for corner to fractional sections 30 and 31, from which A Buckeye, 16 in. dia. bears N. 50 E. 16 links dist. A Hackberry, 15 in. dia. bears S. 79 E. 14 links dist.
	Land, low bottom; subject to inundation 3 or 4 feet deep,
	Timber, Buckeye, Hackberry, Oak, and Hickory.

	North, between sections 29 and 30, Variation 17° 40′ East.
6·50	Enter creek bottom, bearing N. E. and S. W.
13 00	Elk Creek, 200 links wide, runs S. W.
15·00	Enter a small prairie, about 40 acres,
31·00	Leave prairie and enter timber, bearing E. and W.
40·00	Set a post for quarter section corner, from which A Hickory, 14 in. dia. bears N. 78 E. 15 links dist. A Bur Oak, 26 in. dia. bears N. 63 W. 19 links dist.
49·71	A Black Oak, 30 in. dia.

[13]

Township 25 *N. Range* 2 *W. Willamette Meridian.*

CHAINS.	
68·19	A Walnut, 36 in. dia.
80·00	Set a post for corner to sections 19, 20, 29, 30, from which
	A Beech, 15 in. dia. bears N. 24 E. 18 links dist.
	A Blue Ash, 24 in. dia. bears N. 79 W. 10 links dist.
	A Bur Oak, 9 in. dia. bears S. 14 W. 10 links dist.
	A Black Oak, 8 in. dia. bears S. 11 E. 14 links dist.
	Land, first half-mile, level prairie, and brushy, Oak and Hazel; second half-mile, some good timber, Oak, etc.; thick undergrowth, same.
	East, on a *random* line between sections 20 and 29,
	Variation 70° 25′ East,
40·00	Set a post for temporary quarter section corner,
80·10	Intersected the N. and S. line 20 links North of the corner to sections 20, 21, 28, 29, from which corner I run
	North 89° 58′ West, on a *true* line between sections 20 and 29,
	Variation 70° 25′ East,
40·05	Set a post for quarter section corner, from which
	A Sugar-Tree, 24 in. dia. bears N. 17 W. 20 links dist.
	A Walnut, 14 in. dia. bears S. 10 E. 36 links dist.
80·10	The corner to sections 19, 20, 29, 30,
	Land level, and rather wet,
	Timber, Oak, Sugar-Tree, Beech, and Walnut; open woods.
	West, on a *random* line between sections 19 and 30,
	Variation 17° 40′ East,
40·00	Set a post for temporary quarter section corner,
75·53	Intersected the West boundary of the Township 20 links South of the corner to sections 19 and 30, from which corner I run
	South 89° 51′ East, on a *true* line between sections 19 and 30,
	Variation 17° 40′ East,
35·52	Set a post for quarter section corner, from which
	A Sugar-Tree, 18 in. dia. bears N. 26 W. 23 links dist.
	An Ash, 10 in. dia. bears S. 86 E. 32 links dist.
75·52	The corner to sections 19, 20, 29, 30,
	Land level; rich soil; not subject to inundation,
	Timber, Sugar-Tree, Beech, Walnut, and Ash; undergrowth, Spice, Prickly Ash, and vines.
	February 11th, 1854.
	North, between sections 19 and 20,
	Variation 17° 40′ East,
7·70	A Bur Oak, 20 in. dia.
27·16	A Locust, 18 in. dia.
34·00	A pond, 200 links wide, muddy bottom, and low banks; water not so deep as to prevent measuring across on the line with the chain. This pond extends about 15 chains East into section 20, and lies mostly in section 19, extending West,
40·00	Set a post for quarter section corner, from which
	A Beech, 9 in. dia. bears N. 56 E. 44 links dist.
	A Lynn, 12 in. dia. bears S. 36 W. 111 links dist.
49·00	The S. W. bank of a lake to be meandered,

[14]

Township 25 N. Range 2 W. Willamette Meridian.

Chains.	
	Set a post for corner to fractional sections 19 and 20, from which
	A Red Oak, 12 in. dia. bears S. 45 W. 21 links dist.
	A Lynn, 15 in. dia. bears S. 23 E. 24 links dist.
	From this corner offset *West* 7·50 chains to a point; thence *North* on an offset line 24·00 chains to a point; thence *East* 7·50 chains to a point in the line between sections 19 and 20—50 links in advance of lake; thence *South* to N. W. margin of lake, 50 links, where set a post for corner to fractional sections 19 and 20, from which
	A Red Oak, 20 in. dia. bears N. 27 E. 31 links dist.
	A Bur Oak, 15 in. dia. bears N. 36 W. 24 links dist.
	This corner is 72·50 chains *North* of the corner to sections 19, 20, 29 30, and from which I continue the line between sections 19 and 20 *North*, counting the distance from the corner to sections 19, 20, 29, 30,
80·00	Set a post for corner to sections, 17, 18, 19, 20, from which
	A Chestnut, 10 in. dia. bears N. 14 E. 14 links dist.
	A Buckeye, 12 in. dia. bears N. 86 W. 13 links dist.
	A Beech, 20 in. dia. bears S. 13 W. 16 links dist.
	A Buckeye, 20 in. dia. bears S. 27 E. 35 links dist.
	Land level; rich soil, but too wet for cultivation,
	Timber, Oak, Walnut, Buckeye, and Beech; undergrowth, Prickly Ash and vines.
	East, on a *random* line between sections 17 and 20,
	Variation 17° 40′ East,
40·00	Set a post for temporary quarter section corner,
79·90	Intersected N. and S. line 7 links North of post corner to sections 16, 17, 20, 21, from which corner I run
	North 89° 57′ West, on a *true* line between sections 17 and 20,
	Variation 17° 40′ East,
39·95	Set a post near the North bank of the lake for quarter section corner, from which
	A White Oak, 12 in. dia. bears N. 33 E. 19 links dist.
	A White Oak, 15 in. dia. bears S. 16 W. 34 links dist.
	From this corner I run South 150 links to a point on the North bank of the lake, where set a meander corner, from which
	A Red Oak, 15 in. dia. bears N. 21 E. 15 links dist.
	An Ash, 12 in. dia. bears N. 16 W. 12 links dist.
79·90	The corner to sections 17, 18, 19, 20,
	Land level and wet; rich soil,
	Timber, Oak, Ash, Elm, and Beech; undergrowth, same, briers and vines.

Meanders of Island Lake.

Begin at the corner to fractional sections 19 and 20, on the N. W. margin of the lake, and run thence along the N. W. margin thereof, in fractional section 20, as follows, viz.:

N. 79 E. 20·00 chains, thence

N. 84 E. 20·43 " to the meander corner 150 links *South* of the quarter section corner on the line between sections 17 and 20, thence

[15]

Township 25 *N. Range* 2 *W. Willamette Meridian.*

CHAINS.

S. 73 E. 16·00 chains, thence
S. 61 E. 14·00 " "
S. 40½ E. 19·22 chains to the corner to fractional sections 20 and 21, on the N. E. bank of lake, at 52·33 chains. At 18·00 chains on this line cross the mouth of a branch, 30 links wide, coming from N. E.

Begin at the corner to fractional sections 20 and 21, on S. E. bank of lake, at 28·94 chains, and run thence along the Southern bank of said lake in fractional section 20, as follows:

S. 70 W. 20·00 chains, thence
S. 85 W. 23·00 " "
N. 70 W. 12·00 " "
N. 30 W. 18·00* " "
*At 14·50 chains cross outlet to lake, 30 links wide, running W. about 5 chains into pond.
N. 63 W. 20·24 " to the corner to fractional sections 19 and 20, at 49·00 chains; thence in section 19, as follows, viz.:
N. 75 W. 5·00 chains, thence
N. 60 W. 2·00 " "
N. 10 W. 6·00 " "
N. 10 E. 6·00 " "
N. 25 E. 3·00 " "
N. 38¾ E. 8·48 " to the corner to fractional sections 19 and 20 on the bank of lake at 72·50 chains.

This lake has low, wet, brushy banks, and has an island of timber in the middle, which ought to be meandered. Timber, around lake, Ash, Maple, and Red Oak. I cause a flag to be set on the North bank of the island *South* of the meander corner, which is 150 links *South* of the quarter section corner on the line between sections 17 and 20. From the meander corner run a base 7·50 *East* to a point, from which the flag bears S. 45 W. which gives for the distance across the water to the flag on the island 7·50 chains. Set a meander post in the place of the flag, from which a Red Oak, 15 in. dia. bears S. 21 W. 24 links, and an Ash, 10 in. dia. bears S. 25 E. 17 links dist. From the meander post I run around the island as follows:

S. 62 E. 7·50 chains, thence
S. 55 E. 10·00 " "
S. 20 E. 5·00 " "
South 4·00 " "
S. 25 W. 6·00 " "
S. 62 W. 5·00 " "
S. 80 W. 4·00 " "
West 3·50 " "
N. 70 W. 5·00 " "
N. 62 W. 15·00 " "
N. 45 W. 10·00 " "
N. 35 W. 6·00 " "
N. 40 E. 6.50 " "
N. 82 E. 8·00 " "

[16]

Township 25 *N. Range* 2 *W. Willamette Meridian.*

CHAINS.	
	S. 88½ E. 14·20 chains, to the meander corner and place of beginning. This island is well timbered, and is good, dry land. Timber, Oak, Hickory, Beech, and Ash; undergrowth, same and vines.
	The line between sections 18 and 19 will strike the river before reaching the range line, I therefore run it
	West, on a *true* line between sections 18 and 19, Variation 17° 40′ East,
7·9	A Buckeye, 15 in. dia.
16·54	A Locust, 24 in. dia.
28·90	Set a post on the left bank of Chickeeles River for corner to fractional sections 18 and 19, from which A Buckeye, 24 in. dia. bears N. 76 E. 22 links dist. A Hackberry, 16 in. dia. bears S. 24 W. 15 links.
	There is an island in the river opposite this corner. To ascertain the distance on the line between sections 18 and 19 to the island, I send my flagman across the slough, who sets the flag on the S. E. bank of the island, and in the line between sections 18 and 19, from the corner to said sections on the left bank of the river. I run South 260 links to a point from which the flag on the island bears N. 45½ W. which gives for the distance 3 79 chains, to which add 28·90 chains, makes
32·69	To the flag. Set a post in the place of the flag for corner to fractional sections 18 and 19, from which A White Oak. 16 in. dia. bears N. 41 W. 37 links dist. A Bur Oak, 14 in. dia. bears S. 81 W. 16 links dist.
36·52	A White Oak, 20 in. dia.
39·10	A Bur Oak, 16 in. dia.
40·00	Set a post for quarter section corner, from which A White Oak, 15 in. dia. bears N. 15 W. 21 links dist. A Walnut, 20 in. dia. bears S. 21 E. 17 links dist.
45·50	Set a post on the N. W. bank of the island for corner to fractional sections 18 and 19, from which A Hackberry, 10 in. dia. bears N. 85 E. 15 links dist. A Hickory, 15 in. dia. bears S. 51 E. 17 links dist.
	From this corner I meander around the island as follows: In section 19,
	S. 60 W. 10 00 chains, thence
	S. 43 W. 8·00 " "
	South 2·00 " "
	East 2·00 " "
	N. 55 E. 4·00 " "
	N. 60 E. 10 00 " "
	N. 66½ E. 14·15 " to the corner to fractional sections 18 and 19, on the S. E. bank of the island, thence in section 18,
	N. 70 E. 10 00 chains, thence
	N. 75 E. 10·00 " "
	N. 25 E. 4·00 " "
	North 2·50 " "
	West 1·00 " "

[17]

Township 25 N. Range 2 W. Willamette Meridian.

CHAINS.	
	S. 66 W. 2·00 chains, thence
	S. 75 W. 4·00 " "
	S. 80 W. 10·00 " "
	S. 63½ W. 21·10 " to the corner to fractional sections 18 and 19, on the N. W. bank of island, and place of beginning.
	Land, on island and main shore, level and rich; not subject to inundation,
	Timber, Oak, Hickory, Ash and Walnut; undergrowth, same and vines.
	North, between sections 17 and 18, Variation 17° 40′ East.
6·57	A Hickory, 20 in. dia.
10·80	Set a post on the left bank of Chickeeles River for corner to fractional sections 17 and 18, from which A Buckeye, 8 in. dia. bears S. 25 W. 15 links dist. A Hackberry, 10 in. dia. bears S. 61 E. 3 links dist. Monday, February 13th, 1854.
	Meanders of the left bank of Chickeeles River through the Township.
	Begin at the corner to fractional sections 4 and 33, in the North Boundary of the Township and on the left and S. E. bank of the river, and run thence down stream with the meanders of the left bank of said river, in fractional sec tion 4, as follows:

Courses.	Distances.	Remarks.
S. 76 W.	18·50 chs.	
S. 61 W.	10·00	
S. 59 W.	8·30	To the corner to fractional sections 4 and 5; thence in section 5—
S. 54 W.	10·70	
S. 40 W.	5·60	
S. 50 W.	8·50	
S. 37 W.	17.00	
S. 44 W.	22.00	
S. 38 W.	26·72	To the corner to fractional sections 5 and 8; thence in section 8—
S. 21 W.	16·00	
S. 10 W.	13·00	
South	8·50	To the head of rapids.
S. 9 E.	5·00	
S. 17 E.	20·00	
S. 10 E.	12·00	To foot of rapids.
S. 22¼ E.	8·46	To the corner to fractional sections 8 and 17.
		Land, along fractional section 8, high, rich bottom; not subject to inundation.
		The rapids are 37·00 chains long; rocky bottom; estimated fall 10 feet.

[18]

Township 25 N. Range 2 W. Willamette Meridian.

Courses.	Distances.	Remarks.
		Meanders in section 17.
S. 17 E.	15·00	At 5 chains discovered a vein of coal, which appears to be 5 feet thick, and may be readily worked.
S. 8 E.	12·00	
S. 4 W.	22·00	At 3·00 chains the ferry across the river to Williamsburg, on the opposite side of the river.
S. 25 W.	17·00	
S. 78 W.	12·00	
S. 71 W.	9·55	To the corner to fractional sections 17 and 18; thence in section 18—
S. 65 W.	15·00	
S. 73¾ W.	15·93	To the corner to fractional sections 18 and 19.
S. 65 W.	14·00	In section 19.
S. 60 W.	23·00	
S. 42 W.	10 00	
S. 20 W.	10·00	
S. 16½ W.	13·83	☞ At 2 chains cross outlet to pond and lake, 50 links wide, to the corner to fractional sections 19 and 24, on the range line, 32·50 chains North of the corner to sections 19, 30, 24 and 25.
		Begin at the corner to fractional sections 25 and 30, on the range line 1 chain *South* of the quarter section corner on said line, and run thence down stream with the meanders of the left bank of Chickeeles River, in fractional section 30, as follows, viz.:
S. 41 E.	20·00	At 10 chains discovered a fine mineral spring.
S. 49 E.	15·00	Here appear the remains of an Indian village.
S. 42 E.	12·00	
S. 12¾ E.	5·30	To the corner to fractional sections 30 and 31; thence in section 31—
S. 12 E.	10·00	
S. 12 W.	13·50	To mouth of Elk River, 200 links wide; comes from the East.
S. 41 W.	9·00	At 200 links across the creek.
S. 58 W.	11·00	
S. 35 W.	11·00	
S. 20 W.	20·00	At 15 chains mouth of stream, 25 links wide; comes from S. E.
S. 23¾ W.	8·80	To the corner to fractional sections 31 and 36, on the range line, and 8·56 chains North of the corner to sections 1, 6, 31 and 36, or S. W. corner to this Township.
		Land along the left bank of Chickeeles River is level, rich soil, and only a small part subject to inundation.
		Timber, Oak, Hickory, Beech, and Elm; not much undergrowth.
		February 14th, 1854.

[19]

Township 25 *N. Range* 2 *W. Willamette Meridian.*

CHAINS. From the corner to sections 18, 19, 13 and 24, I run East, on a *true* line between sections 18 and 19,

Variation 18° 00′ East,

3·52 A Bur Oak, 20 in. dia.

17·31 A White Oak 15 in. dia.

21·00 Set a post on the right bank of Chickeeles River for corner to fractional sections 18 and 19, from which

A White Oak, 15 in. dia. bears N. 10 E. 31 links dist.

A Black Oak, 20 in. dia. bears S. 80 W. 15 links dist.

From this corner the corner to fractional sections 18 and 19, on the N. W. bank of the island, bears *East.*

To obtain the distance across the river between the two corners, I run (from the corner on right bank) *North* 375 links to a point from which the corner on the island bears S. 68 E. which gives for the distance 9·27 chains.

The length of the line between seetions 18 and 19 is 75·77 chains, the several parts of which being as follows:

East of river and across the island, including 3·79 chains across the slough . . . . 45·50 chains.

Across the river N. W. of island . . . . . . 9·27

West of river . . . . 21·00

Aggregate as above . . . . 75·77

From the corner to fractional sections 19 and 24, on the right bank of Chickeeles River, I run up stream with the right bank of said river in fractional section 19, as follows, viz.:

N. 30 E. 20·00 chains, thence

N. 45¼ E. 15·00 chains to the corner to fractional sections 18 and 19; thence, in section 18,

N. 58 E. 10·00 chains, thence

N. 63 E. 17·00 " "

N. 75¾ E. 32·12 " to a point on the right bank of Chickeeles River *North* of the corner to fractional sections 17 and 18, on the left bank of the river; I here set a post for corner to fractional sections 17 and 18, on North side of river, from which

A Black Oak, 15 in. dia. bears N. 25 E. 21 links dist.

A Black Oak, 20 in. dia. bears N. 27 W. 17 links dist.

To obtain the distance across the river, on the line between sections 17 and 18, I run a base line *West* 430 links to a point from which the post corner to fractional sections 17 and 18, on the left and South bank of the river, bears S. 23 East, which gives for the distance 10·13 chains, to which add 10·80 chains, makes

20·93 To the corner to fractional sections 17 and 18, on the right and North bank of the river.

From the corner to fractional sections 17 and 18, on the right and North bank of Chickeeles River, 20·93 chains North of the corner to sections 17, 18, 19, 20, I run

[20]

Township 25 *N. Range* 2 *W. Willamette Meridian.*

CHAINS.	
	North, between sections 17 and 18, counting the distance from the corner to sections 17, 18, 19, 20,
	Variation 18° East,
22·73	A Black Oak, 20 in. dia.
36·45	Intersected the Southern line of Samuel Williams's claim, where set a post for corner to fractional sections 17 and 18, from which
	A Black Oak, 16 in. dia. bears S. 50 W. 22 links dist.
	A White Oak, 20 in. dia. bears S. 21 E. 31 links dist.
	From this corner I run N. 78 E. along the Southern line of the said claim 20 15 chains to the corner tree on the right bank of Chickeeles River and S. E. corner of said claim; thence down stream, on the right bank of said river, in fractional section 17, as follows:
	S. 16 W. 10·00 chains, thence
	S. 45 W. 10·00 " "
	S. 72 W. 10·30 " to the corner to fractional sections 17 and 18.
	Field notes of the survey of a small island in Chickeeles River lying wholly in section 17.
	Cause the flag to be set on the head of the island at a point bearing S. 45 E. from the Black Oak Tree, the S. E. corner to Samuel Williams's claim; from said corner tree run S. 45 W. 215 links to a point *West* of the flag on the head of the island, which gives for the distance from the corner tree to the flag 215 links. Set a meander post in the place of the flag, from which
	A Bur Oak, 16 in. dia. bears S. 10 W. 15 links dist.
	A White Oak, 12 in. dia. bears S. 15 E. 21 links dist.
	From the meander post I run around the island as follows:
	S. 16 W. 9·00 chains, thence
	S. 45 W. 10·00 " "
	S. 10 W. 2·00 " "
	South 1·50 " to the lower end of island, thence
	East 1·50 chains, thence
	N. 75 E. 4·00 " "
	N. 50 E. 5·00 " "
	N. 30 E. 6·00 " "
	N. 10 E. 6·00 " "
	N. 10 W. 3·00 " "
	N. 73 W. 2·96 " to the meander post and place of beginning.
	This island is well timbered; White and Black Oak, and Hickory; not subject to inundation; undergrowth, same, Spice and vines.
	From the corner to sections 7, 18, 12 and 13, on the range line I run
	East, on a *true* line between sections 7 and 18,
	Variation 18° 00′ East,
7·93	Intersected the western line of Samuel Williams's survey of

[21]

Township 25 *N. Range* 2 *W. Willamette Meridian.*

CHAINS. 640 acres, and at said intersection set a post for corner to fractional sections 7 and 18. from which

A White Oak, 15 in. dia. bears N. 25 W. 15 links dist.
A Black Oak, 20 in. dia. bears S. 34 W. 19 links dist.

From this corner I run

N. 12 W. with the Western line of said Williams's claim 23·23 chains to the N. W. corner thereof.

Land gently rolling.

Timber, Oak and Hickory.

From the corner to fractional sections 17 and 18, in the Southern line of Samuel Williams's survey, and 36·45 chains *North* of the corner to sections 17, 18, 19, 20, I run

North, on a *blank* line passing through Samuel Williams's survey, counting the distance from the corner to said sections 17, 18, 19, 20,

Variation 18° 00′ East,

40·00 Point for quarter section corner *in* Samuel Williams's survey, corner not established,

52·50 The road leading into Williamsburg,

80·00 Set a temporary corner to sections 7, 8, 17, 18, in said Williams's claim,

This line passes through the back part of the town of Williamsburg; but I make no connection with the lines of said town.

North, on a *blank* line between sections 7 and 8,

Variation 18° 00′ East,

12·50 To creek, 30 links wide; runs east, comes from N. W.

38·10 Intersected the North Boundary of Samuel Williams's survey, where set a post for corner to fractional sections 7 and 8, from which

A Black Oak, 10. in. dia. bears N. 10 E. 15 links dist.
A Bur Oak, 15 in. dia. bears N. 16 W. 17 links dist.

From this corner I run N. 78 E. on the North line of said claim, 440 links to the N. E. corner thereof, on the right bank of Chickeeles River,

From the corner of fractional sections 7 and 8 in the North line of Samuel Williams's survey,

North, on a *true* line between sections 7 and 8, counting the distance from the temporary corner to sections 7, 8, 17, 18, within said Williams's survey.

40·00 Set a post for quarter section corner, from which

A Black Oak, 15 in. dia. bears N. 25 E. 16 links dist.
A White Oak, 16 in. dia. bears N. 73 W. 12 links dist.

45·17 A White Oak, 18 in. dia.

63·71 A Bur Oak, 15 in. dia.

80·00 Set a post for corner to sections 5, 6, 7, 8, from which

A Red Oak, 20 in. dia. bears N. 20 E. 40 links dist.
A White Oak, 16 in. dia. bears N. 16 W. 43 links dist.
A Red Oak, 24 in. dia. bears S. 80 W. 39 links dist.
A White Oak, 40 in. dia. bears S. 75 E. 22 links dist.

[22]

Township 25 *N. Range* 2 *W. Willamette Meridian.*

CHAINS.	
	Land gently rolling; good, rich soil, Timber, Oak, Hickory, and Ash. February 16th, 1854.
	East, on a *true* line between sections 5 and 8, Variation 18° 00′ East,
5·16	A White Oak, 15 in dia.
7·41	A Bur Oak, 12 in. dia.
10·50	Set a post on the right bank of Chickeeles River for corner to fractional sections 5 and 8 west of river, from which A Red Oak, 30 in. dia. bears N. 58 W. 5 links dist. A Hickory, 12 in. dia. bears S. 42 W. 5 links dist. From this corner the post corner to fractional sections 5 and 8, on the left bank of the river, bears S. 89 E. From a point 16 links *South* of this corner, and *West* of the corner to fractional sections 5 and 8, on the left and East bank of the river, I run North 454 links to a point from which the corner post on the left bank of the river bears S. 63 E. which gives for the distance across the river 8·91 chains. The length of the line between sections 5 and 8, including the distance across the river, is therefore 80·06 chains, viz.: East of river . . . . 60·65 chains. Across river . . . . 8·91 West of river . . . . 10·50 Total . . . . 80·06
	West, on a *random* line between sections 6 and 7, Variation 18° 00′ East,
25·10	A stream, 25 links wide, gentle current, runs South,
40·00	Set a post for temporary quarter section corner,
56·00	A stream, 15 links wide, runs S. E.
76·26	Intersected the West Boundary 21 links North of the corner to sections 6 and 7, from which corner I run North 89° 51′ E. on a *true* line between sections 6 and 7. Variation 18° 00′ E.
36·26	Set a post for quarter section corner, from which A Black Oak, 16 in. dia. bears N. 15 W. 21 links dist. A White Oak, 40 in dia. bears S. 21 W. 33 links dist.
76·26	The corner to sections 5, 6, 7, 8. Land hilly; 2d rate, Timber, Oak, Sugar-Tree, and Hickory; undergrowth, same and Hazel.
	North, on a *random* line between sections 5 and 6, Variation 18° 00′ East,
20·00	Enter windfall, bearing N. 60 W. and S. 60 E.
35·90	Leave windfall, having same bearings,
40·00	Set a post for temporary quarter section corner,
80·06	Intersected the North Boundary of the Township 24 links East of the corner to sections 5 and 6, from which corner I run

[23]

Township 25 *N. Range* 2 *W. Willamette Meridian.*

CHAINS.	South 10′ East, on a *true* line between sections 5 and 6, Variation 18° 00′ East,
40 06	Set a post for quarter section corner, from which A Hickory, 20 in dia. bears N. 18 E. 27 links dist. A White Oak, 24 in. dia. bears S. 31 W. 18 links dist.
80·06	The corner to sections 5, 6, 7, 8. Land rolling, and 2d rate, Timber, Oak, Hickory, Sugar-Tree, and Ash; undergrowth, same and Hazel. February 18th, 1854.

GENERAL DESCRIPTION.

The quality of the land in this Township is considerably above the common average. There is a very fair proportion of rich bottom land, chiefly situated on both sides of Chickeeles River, which is navigable through the Township for steamboats of light draft, except over the rapids in section 8. These rapids are 37 chains long; estimated fall about 10 feet.

The uplands are generally rolling, good 1st and 2d rate land, and well adapted for cultivation. Elk River is a beautiful stream of clear water, running through the Southern part of the Township, and emptying into Chickeeles River, in section 31. There is a fine mill-seat on this stream in section 22.

Timber, chiefly Oak, Beech, Hickory, Hackberry, and Sugar-Tree, and is very equally distributed over the Township, except in the prairie embracing parts of sections 3, 4, 9, 10, 15, and 16.

The town of Williamsburg was laid out by Samuel Williams, some two years since, on the right bank of Chickeeles River, a little below the foot of the rapids. It now contains sixteen houses, and others are being built; has a good landing in front, with a ferry, and has the appearance of thrift and prosperity.

There are several good quarries of stone (principally lime) along the Chickeeles and Elk Rivers, which will afford inexhaustible quantities of excellent building materials. On the line between sections 1 and 12, I discovered gold dust and auriferous quartz, and in section 17, on the left bank of Chickeeles River, opposite Williamsburg, a valuable coal bank. There are three settlements—one on the N. W. quarter of section 10, one on the N. W. quarter of section 15 and N. E. quarter of section 16, and the other on the N. E. quarter of section 23 and N. W. quarter of section 24.

A valuable salt spring was discovered crossing the South Boundary of section 31, running N. W.; also the remains of an Indian village on the left bank of Chickeeles River, in section 30. Fossil remains on the West bank of a small lake in section 26, and ancient works on the l[illegible] bank of Elk River, in the N. E. quarter of section 27

LIST OF NAMES.

A list of the names of the individuals employed to assist in running, measuring, or marking the lines and corners described in the foregoing field notes of Township No. 25 North of the base line of Range No. 2 West of the Willamette meridian, showing the respective capacities in which they acted:

PETER LONG, Chainman.
JOHN SHORT, Chainman.
GEORGE SHARP, Axman.
ADAM DULL, Axman.
HENRY FLAGG, Compassman.

We hereby certify that we assisted Robert Acres, deputy surveyor, in surveying the exterior boundaries and subdividing Township number twenty-five North of the base line of Range number two west of the Willamette meridian, and that said Township has been in all respects, to the best of our knowledge and belief, well and faithfully surveyed, and the boundary monuments planted according to the instructions furnished by the Surveyor-General.

PETER LONG, *Chainman.*
JOHN SHORT, *Chainman.*
GEORGE SHARP, *Axman.*
ADAM DULL, *Axman.*
HENRY FLAGG, *Compassman.*

Subscribed and sworn to by the above-named persons, before me, a Justice of the Peace for the county of , in the State [or Territory] of , this day of , 186 .

HENRY DOOLITTLE,
Justice of the Peace.

I, Robert Acres, deputy surveyor, do solemnly swear that, in pursuance of a contract with surveyor of the public lands of the Unitted States in the State [or Territory] of , bearing date the day of 186 , and in strict conformity to the laws of the United States and the instructions furnished by the said Surveyor-General, I have faithfully surveyed the exterior boundaries [or subdivision and meanders, as the case may be] of Township number twenty-five North of the base line of Range number two West of the Willamette meridian, in the aforesaid, and do further solemnly swear that the foregoing are the true and original field notes of such survey.

ROBERT ACRES,
Deputy Surveyor.

Subscribed by said Robert Acres, deputy surveyor, and sworn to before me, a Justice of the Peace for county, in the State [or Territory] of , , this day of , 186 .

HENRY DOOLITTLE,
Justice of the Peace.

To each of the original field books, the Surveyor-General will append his official approval, according to the following form, or so varied as to suit the facts in the case:

SURVEYOR-GENERAL'S OFFICE AT

186 .

The foregoing field notes of the survey of [here describe the survey], executed by Robert Acres, under his contract of the day of , 186 , in the month of , 186 , having been critically examined, the necessary corrections and explanations made, the said field notes, and the surveys they describe, are hereby approved.

A. B.
Surveyor-General.

To the copies of the field notes transmitted to the seat of government, the Surveyor-General will append to each township the following certificate:

I certify that the foregoing transcript of the field notes of the survey of the [here describe the character of the surveys, whether meridian, base line, standard parallel, exterior township lines, or subdivision lines, and meanders of a particular township] in the State [or Territory] of , has been correctly copied from the original notes on file in this office.

A. B
Surveyor-General.

DIAGRAM.

Town. No. 25 N. Range No. 2 W. Willamette Meridian.

THE BOUNDARIES OF THE PUBLIC SURVEYS NOT TO BE CHANGED.

Congress, as early as the year 1805, laid down certain general principles in regard to the *unchangeableness* of the lines and corners established by government surveyors, which have continued operative down to the present time, and are still in full force. These principles are contained in the second section of an act entitled "An act concerning the mode of Surveying the Public Lands of the United States," approved February 11th, 1805, and are as follows, to wit:

1st. "*All the* CORNERS *marked in the surveys returned by the Surveyor-General, shall be established as the proper corners of sections or subdivisions of sections which they were intended to designate;* and the corners of half and quarter sections, not marked on said surveys, shall be placed as nearly as possible equidistant from those two corners which stand on the same line."

2d. "*The* BOUNDARY LINES *actually run and marked in the surveys returned by the Surveyor-General, shall be established as the proper boundary lines of the sections or subdivisions for which they were intended;* and the length of such lines as returned by the Surveyor-General aforesaid, shall be held and considered as the true length thereof."

Experience has demonstrated the wisdom of this enactment; no law ever passed by Congress has contributed so much to prevent disputes in regard to boundaries of the public lands. Considering the extent of territory over which the public surveys have been extended, embracing whole states now thickly settled with people, and affecting interests involving many thousands of dollars, cases of litigation growing out of disputed boundaries are surprisingly rare.

The law referred to enunciates the unvarying rule, that

all corners of the public surveys marked in the field and duly returned by the Surveyor-General, shall stand as the true corners which they were intended to designate. It will be understood that only such boundaries as are established by the Surveyor-General or the deputy surveyor, in the line of their official duties, and in pursuance of law, come under this rule. If, for instance, the Surveyor-General were to return a section divided into irregular tracts not recognized by the laws governing the survey of the public lands, such boundaries would not stand.

A case of this description in the State of Alabama was decided by the Supreme Court of the United States in 1845. (3d Howard, pp. 650–73.) In this case the Surveyor-General returned the plat of a fractional section divided into two lots of unusual form and unrecognized areas, instead of returning one full quarter section and remaining fractions, as required by the law of 1820. The decision of the court was as follows:

"There is nothing in any of the acts of Congress to authorize the division made by the Surveyor-General, and it being a violation of the laws and contrary to the duties of his office, it must be regarded as void."

But where the corners are established by the proper officer in pursuance of the system of subdivision authorized by law, they must be regarded as the true corners which they represent. Even if it is subsequently found that the post is out of line, or that the intervals are unequal or incorrect; no party has a right to correct such errors except the general government, and it possesses the power only while the title to the lands affected by the change is yet in the United States. After the lands have passed into the hands of private parties, the government lines and corners, as marked in the field, must govern in determining the boundaries of all legal subdivisions, when they can be found and identified; when they are *missing*, recourse must be had to the official plats and field notes of the government survey. The foregoing remarks apply as well to the *lines* as to the *corners* of the public surveys; neither must be changed except by the government agent, under the circumstances mentioned above.

Although this rule may in some instances work a hardship to individuals, giving to one party a larger part and to another a smaller, yet it is one of the conditions under which the parties acquire their land, and the evil in these exceptional cases is immeasurably overbalanced by the advantages of having fixed and unchangeable boundaries to the public lands.

RESTORING OBLITERATED BOUNDARIES.

When extinct lines or corners of the public lands are required to be re-established, a county surveyor or other competent person is usually employed by private parties. It is not the province of the General Land Office to direct the operations of any but government surveyors engaged in the public service; yet, obliterated boundaries must be restored in conformity with the laws and regulations under which they were originally established, and the General Land Office, which is the executive office of the land department, and from which all rules and regulations relative to surveying under the U. S. laws emanate, is the most competent source from which to derive such knowledge as will enable the surveyor to perform his work in a proper manner, and instructions from this source carry with them great weight in the settlement of disputed boundaries.

Information of this character has from time to time been imparted in response to the numerous calls of private surveyors; and the directions which follow are based upon the instructions and correspondence of the office upon the subject, running through many years, and are, it is believed, sufficiently full and explicit to meet all ordinary cases.

If one uniform method had always been employed in establishing the original corners of the public surveys, it would be an easy matter to lay down rules for restoring such of them as from time to time become obliterated. But the system of surveying has undergone modifications at different periods and in various localities, which make it impracticable to give fixed rules that will apply in all cases.

For instance, the mode of surveying the exterior and subdivisional lines of a township at one period resulted in establishing *two sets* of section corners on the township boundary; at a still earlier day *three sets* were sometimes established on the *range lines;* and the system now in practice requires that the section lines shall close on the corners previously established on the township exteriors, thus making but *one set* of corners, except on base lines and standard parallels, where "double corners" are established.

To restore extinct boundaries of the public lands correctly, the surveyor must have some knowledge of the manner in which townships were subdivided by these several methods. Without this knowledge he may be greatly embarrassed in the field, and is liable to mistake the corner of one section for that of another, and thereby be led into error. The following brief explanation of the modes which have been practiced, will be of service to surveyors who may be called upon to restore obliterated boundaries of the public surveys.

Where two sets of corners were established on township boundaries, one set was planted at the time the exteriors were run, those corners on the *north* boundary belonging to the sections and quarter sections *north of said line*, and those on the *west* boundary belonging to the sections and quarter sections *west of the line.* The other set of corners was established when the township was subdivided. The mode of proceeding was as follows: The north-and-south section lines were run *from* the corners pre-established on the south boundaries *due north*, and at the points of intersection with the north boundaries other section corners were erected. The east-and-west section lines were closed on the corners previously established on the east boundary, but were run on a *due west* course from the last interior section corner to the range line, and new section corners were planted thereon at the points of intersection. It will be seen that this method resulted in establishing two sets of corners on all four sides of the townships.

When three sets of corners were established on the *range lines*, the subdivisional surveys were made in the same manner as described above, except that the east-and-west lines, instead of being closed on the corners pre-established on the east boundary, were run *due east* from the last interior section corner, and new corners were erected at the points of intersection with the range line.

The method of subdividing now in practice requires that the section lines be run *from* the corners on the *south boundary* and close on the existing corners on the east, north, and west township boundaries, except where said boundary is a base line or standard parallel. This method is described in detail in the preceding pages of the surveying Manual proper.

With these brief preliminary explanations, we will proceed to consider the

MODE OF RESTORING LOST CORNERS.

From what has been said in regard to the *unchangeableness* of the government surveys, it will be plain that extinct lines or corners must be restored to the exact locality they originally occupied, if possible. It follows, therefore, that resort should first be had to the marks in the field. The surveyor should first seek to identify the missing corner on the ground, by the aid of the bearing trees or witness mounds, line trees, etc., described in the original field notes. When two or more witness trees or mounds can be found, they afford the best means for restoring a missing corner to its original position that can be had. If the corner cannot be identified in this manner, clear and unquestionable testimony as to the locality it originally occupied should be taken, if such testimony be obtainable.

After all the rules and instructions that can be given for re-establishing obliterated public surveys, much will depend upon the skill, fidelity, and good judgment of the surveyor for the correct performance of the work. The most difficult point in laying down instructions, and one on which something must be left to the good judgment of the surveyor, is in regard to what shall be considered *sufficient evidence* in these cases. A definite rule can no more be prescribed in this respect concerning surveying, than a law could be enacted defining just how strong the testimony in a given case should be to satisfy the mind of a justice or juryman. The sound judgment of a competent surveyor in this matter will seldom lead him into error.

PROPORTIONATE MEASUREMENTS.

In retracing lines it frequently happens that the measurements do not agree with those stated in the government field notes. This discrepancy generally arises from a difference in the length of the respective chains used, or a want of proper care in straightening and leveling the chain, or in sticking the pins, on the part of one set of chainmen or the other, but is sometimes owing to an error in tallying committed by the government chainman. When these differences in measurement occur, the county surveyor must *in all cases* establish his corners at intervals PROPORTIONATE *to those given in the government field notes.* This rule must be observed even if the original interval be one or more tallies too many or too few.

EXTINCT INTERIOR SECTION CORNERS.

To restore a missing section corner in the interior of the township.

Run a right line between the nearest noted station trees or well-defined corners, north-and-south and east-and-west of the lost corner; and at the point of intersection of the lines thus run, plant the section corner, with new bearings and distances from it to the nearest durable objects.*

EXTINCT SECTION CORNERS ON TOWNSHIP LINES.

1. *When the corner on a township line is common to four sections.* In this case the corner is common to two sections in one township and two in another, and should be restored in accordance with the directions for re-establishing interior section corners.

2. *Where double section corners were originally established, one of which is still standing and it is required to restore the other.* It will be borne in mind that the corners established when the exterior lines were run, belong to the sections in the township north and west of those lines respectively. It must therefore first be determined beyond a reasonable doubt to which sections the existing corner belongs. By testing the courses and distances to witness trees recorded in the field notes, and remeasuring given distances from known corners, the surveyor will be enabled to decide this question correctly.

Having ascertained to which township the existing

---

* But the proportional distance E. N. W. and S. must be preserved.

corner belongs, the missing corresponding corner of the section in the opposite township may be re-established in line, north or south of the existing corner, as the case may be, at the distance stated in the field notes.

This mode is considered preferable to that of retracing the section line, because these double corners are not usually more than one or two chains apart, and this distance can be measured with greater accuracy than the section line could be re-run.

[NOTE.—The surveyor should not, however, content himself with simply chaining the distance between the double corners, but should *in all cases* test his work by coursing on trial lines and by chaining to known corners. It may be remarked here, and the remark applies to all cases of rechaining referred to in these instructions, that the chain used should be carefully compared with that used by the United States deputy surveyor, by rechaining stated distances between noted objects or corners in the original survey, and where it is found to be either longer or shorter than the chain originally employed, allowance must be made therefor, keeping in mind that all distances in the resurvey must be made *proportional* to those of the original measurements.]

This note is to be taken in connection with, and considered a part of, the instructions which follow for restoring extinct corners.

3. *Where both corners are missing, and it is required to restore the one established when the township line was run.*

Run a straight line between the nearest noted station trees or corners north and south, or east and west, as the case may be, and plant the missing corner at the point in such line indicated by the distances from said trees or corners given in the field notes. The measurements of the survey of the exterior lines should govern, and if the chaining of the surveyor does not agree with the original measurements, the difference should be divided proportionally between the respective distances remeasured, as directed in the preceding note.

The restored corner will be common to two sections either north or west of the township boundary; and the section line north or west, as the case may be, from said corner should also be retraced, to test the accuracy of the result.

4. *Where both corners are missing, and it is required to restore the one established when the township was subdivided.*

Retrace the section line which closed on the missing corner, and plant the section post at the intersection with the township boundary line. Test the result by measuring the distances to noted objects on the township line, and comparing the measurements with those given in the original field notes. The restored corner will, of course, be common to two sections south or east of the township line.

5. *When triple corners have been established on range lines, one or two of which have become obliterated, and it is required to restore either of them.*

It will be remembered that only *two* of these corners are actual corners of sections, those established when the range line was run not corresponding with the boundaries of the sections either east or west of said line. The surveyor will, therefore, first proceed to identify the existing corner or corners, and then plant the missing corner in line north or south of them, according to the distances stated in the original field notes, and test the correctness of the result as heretofore directed.

If the distances between the triple corners are not stated in the field notes, the required corner must be restored by retracing the section line closing on said corner, as directed in the case of double corners similarly situated.

6. *Where triple corners have been established, all of which are missing.*

The required corner should be restored in accordance with the instructions for re-establishing extinct *double corners* similarly situated: *i.e.* by retracing the section line which closed on the missing corner, either east or west. The range line should also be rechained north and south to the nearest corner, to make sure the correctness of the result.

EXTINCT QUARTER SECTION CORNERS.

Quarter section corners, except those on section lines which close on the north or west boundaries of townships, are required to be established *equidistant* between the section corners; it is an easy matter, therefore, to restore such of them as may become extinct, if the true section corners be known.

1. *To restore lost quarter section corners on a township boundary.*

Only one set of quarter section corners are actually marked in the field on township lines, and they are established when the exteriors are run. When there are double section corners, the quarter posts are considered as standing midway between the corners of their respective sections, and when required to be marked in the field, should be so placed. This is also true in regard to triple corners; but great care must be taken not to mistake the corner of one section for that of another.

2. *Quarter section corners on section lines which close on the north and west township boundaries.*

These corners must be re-established according to the original measurement, at forty chains from the last interior section corner. If the measurements do not coincide with the original survey, the excess or deficiency must be proportionally divided between the two distances as stated in the field notes.

For example, suppose the line between sections 5 and 6, or 6 and 7, to be 81·25 ch. according to the field notes, but according to present measurement it is 80·85 ch. Then, as 81·25 : 80·85 :: 40=39·75$\frac{3}{10}$ ch. The quarter section corner must therefore be placed at 39 chains and 75$\frac{3}{10}$ links from the last interior section corner, and consequently 41 ch. and 09$\frac{7}{10}$ links from the township boundary, according to present measurement. The same course should be pursued in the case of *anomalous* sections, where two or more corners occur at intervals of 20 chains, between the regular half-mile post and the township line.

3. *Interior quarter section corners, except in the preceding cases*, must be re-established equidistant between the section corners, and in a right line between the nearest noted station trees or section corners on either side of them.

EXTINCT TOWNSHIP CORNERS.

1. *When the missing township corner is common to four townships.*

These township corners should be restored in accordance with the directions for re-establishing interior section corners.

2. *When the missing corner is on a correction line and common to only two townships.*

Both the correction line and the range line of the township to which the required corner belong, should be carefully retraced and measured to a sufficient distance to insure accuracy in the courses and distances; and if the range line should not intersect the correction parallel at the distance from the nearest township corner stated in the field notes, the work should be further examined and rigidly tested.

EXTINCT MEANDER CORNERS.

1. These corners should be restored by retracing the lines which closed upon them in the direction that they were run by the government surveyor.

Fractional section lines closing on Indian boundary lines, reservations, private grants, etc. should be retraced in the same manner.

2. It may not unfrequently happen that, after proceeding to restore a lost corner in the manner described in the foregoing pages, the surveyor may come upon some traces of the original corner; when such is the case, and the traces unmistakably indicate the original locality of the corner, the resurvey must of course be made to conform therewith.

EARLY LAWS OF CONGRESS RELATING TO SURVEYING THE PUBLIC LANDS.

The following brief synopsis of some of the early surveying laws may be of service to county surveyors in the regions of the old surveys.

The first enactment of the National Legislature in regard to surveying the public lands was an ordinance passed by the Congress of the Confederation, May 20th, 1785, which prescribed the mode for the survey of the "Western Territory." This ordinance provided that said territory should be divided "into townships of six miles square, by lines running due north and south, and others crossing them at right angles," as near as might be.

It was further provided that "the first line running north and south should begin on the Ohio river, at a point due north from the western termination of a line run as the southern boundary of the State of Pennsylvania, and the first line running east and west should begin at the same point, and extend through the whole Territory."

The townships were to be designated by numbers from south to north, beginning each range with number one; and the ranges were to be distinguished by numbers from east to west, the first range extending from the Ohio river to Lake Erie. The first seven of these series of townships constitute what was called "the seven ranges."

The first law passed by the Federal Congress in regard to surveying the public domain was approved May 18th, 1796, and applied to "the territory northwest of the river Ohio and above the mouth of the Kentucky river."

The second section of said act provided for dividing such lands as had not already been surveyed or disposed of, "by north and south lines run according to the true meridian, and by others crossing them at right angles, so as to form townships of six miles square," etc. It was also provided that "one-half of said townships, taking them alternately, should be subdivided into sections containing, as nearly as may be, 640 acres each, by running parallel lines through the same each way at the end of every two miles, and marking a corner on each of said lines at the end of every mile."

An act amendatory of the above, approved May 10th, 1800, directed that "the interior lines of townships intersected by the Muskingum, and of all the townships lying east of that river, which had not before been actually subdivided into sections, should also be run and marked in the manner prescribed by the said act for running and marking the interior lines of townships directed to be sold in sections of six hundred and forty acres each." And in all cases where the exterior lines of the townships thus to be subdivided exceeded or fell short of six miles, the excess or deficiency was to be added to or deducted from the western or northern tier of sections. Said act also provided that the northern and western tiers of sections should be sold as containing only the quantity expressed on the plats, and all others as containing the complete legal quantity.

The two acts last named form the original basis of the present system of subdivisional surveys, as illustrated in succeeding pages. Sundry modifications and additions have from time to time been incorporated by subsequent acts of Congress.

An act "regulating the grants of land appropriated for military purposes," etc., approved June 1st, 1796, provided for dividing the "Virginia military tract" in the State of Ohio into townships of *five* miles square, each to be subdivided into quarter townships containing 4000 acres. By the 6th section of an act amendatory of said act, approved March 1st, 1800, the Secretary of the Treasury was authorized to subdivide said quarter townships (then called sections) into lots of 100 acres each, bounded as near as practicable by parallel lines, 160 perches in length and 100 perches in width.

It will be borne in mind that these subdivisions were made by *protraction* merely, upon the plats in the office of the Secretary of the Treasury; and afterwards, when a sufficient number of lots to warrant it had been located, an actual survey was made in the field. In some instances, when these lots came to be surveyed, they could not be made to correspond with the plats; fractional lots, which appeared on the plats, were crowded entirely out of the township by actual survey. A knowledge of this fact will explain difficulties which are sometimes met with in the districts thus divided.

The 1st section of the act, approved March 26th, 1804, made it the duty of the Surveyor-General to cause the public lands north of the river Ohio, and east of the river Mississippi, to be surveyed in townships six miles square, and divided in the same manner as provided by law in relation to the lands northwest of the river Ohio and above the mouth of the Kentucky river.

THE BOUNDARIES AND CONTENTS OF SUBDIVISIONS.

The extent of the public domain to be divided into suitable tracts for settlers rendered some fixed laws necessary in regard to the boundaries and contents of the subdivisions. This subject early engaged the attention of Congress, and the following law relating thereto was passed and approved February 11th, 1805, to wit:

"AN ACT concerning the mode of surveying the public lands of the United States.

"SEC. 2. *And be it further enacted*, That the boundaries and contents of the several sections, half sections, and quarter sections of the public lands of the United States, shall be ascertained in conformity with the following principles, any act or acts to the contrary notwithstanding:

"1. All the corners marked in the surveys returned by the Surveyor-General, or by the surveyor of the land south of the State of Tennessee, respectively, shall be established as the proper corners of sections, or subdivisions of sections which they were intended to designate; and the corners of half and quarter sections not marked on the said surveys, shall be placed as nearly as possible equidistant from those two corners which stand on the same line.

"2. The boundary lines actually run and marked in the surveys returned by the Surveyor-General, or by the surveyor of the land south of the State of Tennessee, respectively, shall be established as the proper boundary lines of the sections, or subdivisions, for which they were intended; and the length of such lines, as returned by either of the surveyors aforesaid, shall be held and considered as the true length thereof. And the boundary lines which shall not have been actually run and marked as aforesaid, shall be ascertained by running straight lines from the established corners to the opposite corresponding corners; but in those portions of the fractional townships where no such opposite corresponding corners have been or can be fixed, the said boundary lines shall be ascertained by running from the established corners due north and south, or east and west, lines, as the case may be, to the water-course, Indian boundary line, or other external boundary of such fractional township.

"3. Each section, or subdivision of section, the contents whereof shall have been, or by virtue of the first section of this act, shall be returned by the Surveyor-General, or by the surveyor of the public lands south of the State of Tennessee, respectively, shall be held and considered as containing the exact quantity expressed in such return or returns; and the half sections and quarter sections, the contents whereof shall not have been thus returned, shall be held and considered as containing the one-half or the one-fourth part respectively of the returned contents of the section of which they make part."

By "An act providing for the division of certain quarter sections in future sales of the public lands," approved February 22d, 1817, it is provided that from and after the first day of September of that year, "in every case of the division of a quarter section" (of the sections designated by numbers 2, 5, 20, 30, and 33), "the partitions shall be made by a line running due north and south." In all other respects the same laws to prevail as in the case of other public lands.

"An act making further provision for the sale of the public lands," approved April 24th, 1820, provides that from and after the first day of July following, "in every case of the division of a quarter section, the line for the division thereof shall run north and south, and the corners and contents of half quarter sections which may thereafter be sold, shall be ascertained in the manner and on the principles directed and prescribed by the second section of the act of February 11th, 1805."

"An act supplemental to the several laws for the sale of the public lands," approved April 5th, 1832, provides that from and after the first day of May following, "in every case of a division of a half quarter section" (in all the public lands of the United States) "the line for the division thereof shall run east and west, and the corners and contents of quarter-quarter sections, which may thereafter be sold, shall be ascertained, as nearly as may be, in the manner and on the principles directed and prescribed by the 2d section of the act of February 11th, 1805; and fractional sections containing fewer, or more than 160 acres, shall in like manner, as nearly as may be practicable, be subdivided into quarter-quarter sections, under such rules and regulations as may be prescribed by the Secretary of the Treasury."

The act of February 11th, 1805, has never been suspended or repealed, and consequently is still in full force. It annulled all previous acts or parts of acts in conflict with its provisions. No single act regarding the public lands ever passed by Congress is equal in importance to this, and yet it is imperfecly understood in the land states. From this law we derive the rules for subdividing sections laid down in these pages.

SUBDIVIDING OF SECTIONS.

Only the exterior lines of sections are actually run and marked in the field by the United States deputy, and the only corner boundaries of legal subdivisions established by him are the section corners and intermediate half-mile corners on the exterior of the section.

Sections are divisible into quarters containing 160 acres each, half-quarters containing 80 acres, and quarter-quarters of 40 acres each, and government disposes of the public lands by these subdivisions; but when purchasers or settlers desire to have the boundaries of these tracts established and marked on the ground, it is done by a county or other private surveyor.

The manner of subdividing sections has been the subject of much perplexity to surveyors, and even to state legislators. The original corners established by the government surveyor must form the basis of all such subdivisions, and as these corners were sometimes established out of their proper positions, various plans for subdividing sections in such cases have been devised, with the view to correct the error in the original survey.

The legislatures of at least three of the public land states have passed laws upon this subject, each differing from the others, and all in conflict with the law of Congress. These laws must, in time, prove a fruitful source of litigation and vexatious annoyance to the land owners, and ought to be corrected as speedily as possible, and made to conform to the United States law.

It must be borne in mind that the boundaries of the public lands established by the government surveyors are *unchangeable*. This subject is treated of more fully in a preceding chapter, to which reference is made in this connection. The following rules will enable the surveyor to divide the section into its legal subdivisions in conformity with the law of Congress:

1. Wherever one or more of the original corners of a section was established out of place, the area of every legal subdivision in said section is affected thereby; *i.e.* some of the subdivisions will contain more than the regular quantity, and others will contain less. It will be useless for the surveyor, therefore, when called upon to subdivide a section where one of the original corners was established out of line or out of measure, to attempt to make such a division as will give an equal area to even two of the subdivisions; it cannot be done without violating the rules prescribed by Congress in such cases.

2. The original section and quarter section corners established by the government surveyor must stand as the true corners which they were intended to represent. This is true whether the corners be in place or not.

3. The quarter-quarter corners not established by the deputy surveyor must be planted equidistant *and on line between the quarter post and the section corner.*

4. All the subdivisional lines of a section must be *straight lines running from the proper corner in one exterior line to its corresponding corner in the opposite boundary of the section.* THERE IS NO EXCEPTION TO THIS RULE.

5. In fractional sections where no opposite corresponding corner has been or can be established, any required subdivision line of such section must be run from the proper original corner in the boundary line, due east and west, or north and south, as the case may be, to the watercourse, Indian reservation or other exterior boundary of said fractional section.

It is evident from the peculiar language of the law, that if the object which made the section fractional at the time the original survey was made shall subsequently have been removed so as to admit of establishing the omitted corner, said corner should first be planted in accordance with the principles laid down in the preceding pages, and then, instead of running the subdivision line on a due north or west course as above directed, run a *straight* line between the corner so established, and the corresponding corner in the opposite boundary of the section.

Anomalous Sections. It has sometimes happened in finishing up the public surveys that sections longer than one mile have been made in order to close on to some previously established boundary line. These are denominated "Anomalous Sections," and the same principle governs the establishment of the corners by the government surveyor in such cases that applies to the establishing of corners in the north and west tiers of sections in a township.

Fig. 3.

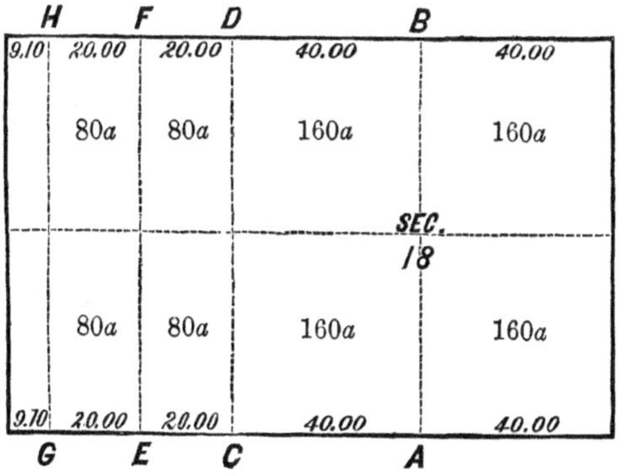

For example, suppose the anomalous section be 129 chs. 10 lks. from East to West. (See Fig. 3.) The posts *A B C* and *D* must be established just 40 and 80 chains, respectively, west from the section corners, and posts *E F G H* must be established at every 20 chains of the remaining distance west to the boundary line.

Anomalous sections should be subdivided by running straight lines from the respective corners on the south boundary to their opposite corresponding corners, the same as in regular sections.

We publish several letters of the General Land Office containing instructions for restoring lost corners and subdividing sections in various cases, from which it will be seen that the foregoing rules and instructions upon this subject are in harmony with the views of this office.

The following examples will illustrate some of the more difficult of the rules for restoring corners.

FIG. 1.

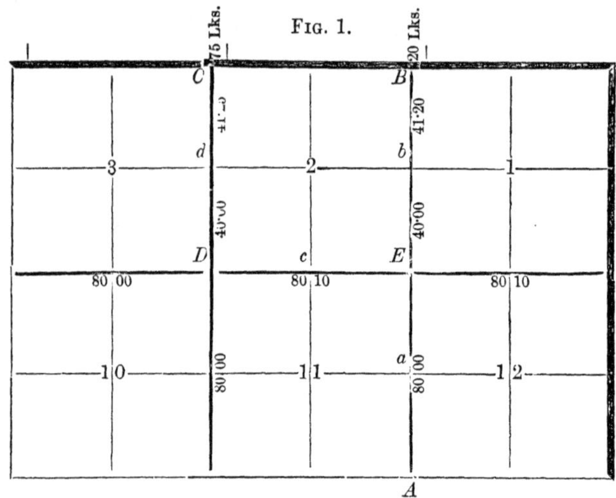

Example 1.—Required to restore the missing section corners *B*, *C*, *D*, *E*, and the quarter section corners *a*, *b*, *c*, *d*. Fig. 1.

Mode.—In this example it will be observed that two sets of section corners were established on the north boundary of the township. From the original field notes it appears that the section line between sections 1 and 2 intersected the township line 20 links west of the corner established when said township line was run. Therefore plant the corner *B* 20 links west of such corner; then proceed to the corner *A* and run a random line north, setting temporary posts at 40 chains, 80 chains, and 120 chains, noting the excess or deficiency on the last half mile, and the falling east or west of the newly erected corner.

Calculate a course that will run a true line from *B* to *A*, and if the distance by the present measurement be more or less than that stated in the original field notes, plant the permanent quarter section corner *b* at a proportional distance:

Thus; the original distance from *A* to *B* was 161 chains and 20 links: suppose the distance by the resurvey to be 162 chains 10 links, then—

As 161·20 : 162·10 :: 41·20 (*B* to *b*, original measurement) = 41·43. The permanent quarter section corner *b* should therefore be re-established 41 chains 43 links south from *B*.

The distance by present measurement from *b* to *A* will of course be (162·10 — 41·43 =) 120 chains and 67 links. The remaining corners *B* and *a* must likewise be restored at proportionate intervals, but as these intervals were equal in the original survey they will be equal in the resurvey; therefore, by dividing the remaining distance into three equal parts we shall have the correct distance at which these corners should be re-established, to wit: 120·67 ÷ 3 = 40·22½ = the true distance from *b* to *B* and also from *E* to *a* and *a* to *A*.

Proceed in the same manner to restore the corners *C*, *D*, *d*, after which plant the quarter section corner *c* equidistant between the section corners *D E*.

In the foregoing example it will be observed that the boundary lines of the sections are of uniform length east and west. Were these lines of various lengths, say from 5 to 75 links, the mode described in the preceding example would not be correct, for it will be remembered that *all* corners are to be restored at *proportional* distances. This is as true in regard to east and west lines as it is of north and south lines; there is no difference in this respect. When, therefore, there is any material difference in the intervals between the section corners east and west, as is frequently the case, the distances must be made proportional east and west as well as north and south, even though at the expense of regularity in the configuration. The mode of proceeding in such cases will be better understood by reference to the following illustration:

Fig. 2.

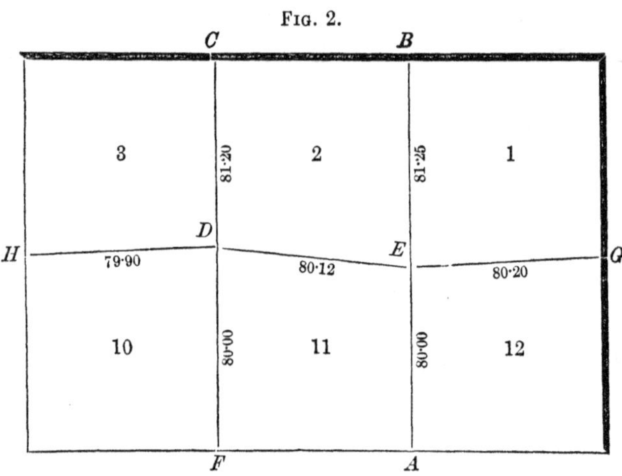

Example 2.—Required to restore the section corners *D* and *E* and all the quarter section corners. Fig. 2.

Mode.—It appears from the field notes of the United States survey that the original intervals between these *section* corners were unequal, therefore they cannot be restored equidistant, but the proportional distances must be preserved. To this end the county surveyor should remeasure the section lines *A B*, *C F*, and *G H*. This done he will have the data from which to make the necessary calculations to enable him to re-establish the lines and corners correctly.

Suppose the result of the remeasurements to be as follows: *A B*, 162·20 chains; *C F*, 160·80 chains; and *G H*, 242·40 chains. Now the distance from *A* to *B* is set down in the original field notes at 161 chains 25 links. Therefore,

As 161·25 : 162·20 : : 80·00 = 80·47 = the true length of the line *A E*, according to present measurement, and 162·20 — 80·47 = 81·73 chs. from *E* to *B*. So also—

As 161·20 : 160·80 : : 80·00 : 79·80 chs. *F* to *D*.

And 160·80 — 79·80 = 81·00 chs. *D* to *C*.

The distance from *G* to *H* as indicated in the original field notes, was 240·22 chs. Then,

As 240·22 : 242·40 : : 80·20 : 80·92¾ chs. *E* to *G*.

240·22 : 242·40 : : 80·12 : 80·84 chs. *D* to *E*.

And 242·40 — (80·92¾ + 80·84 =) 161·76¾ = 80·63¼ chs. *D* to *H*.

Having provided the above data, proceed to *A* and re-measure the south boundary of section 12. Having calculated a course that will run from *A* to *E*, plant the ¼ section corner at 40·86½ chs., and the section corner *E* at 81·73 chs. Then run a random line to *G*, planting the ¼ section corner at 40·46⅜ chs., and correct back on a true line.

The original distance from *E* to the ¼ section corner north of it was, of course, 40 chs. The distance from *E* to *B*, by present measurement, is 81·73 chs. Then,

As 81·25 : 81·73 :: 40·00 : 40·23½.

Calculate a course which will run from *E* to *B*, and establish the quarter section corner 40·23½ chs. north from *E*.

Return to *F* and proceed in the same manner to restore the corners on the section line from *F* to *C*.

By this mode, the quarter section corners between *E B* and *D C* will have been established at proportionate distances between the respective section corners, and all the other ¼ section corners *equidistant* between their respective section corners, in conformity with the law.

GENERAL LAND OFFICE LETTERS.

DEPARTMENT OF THE INTERIOR,
GENERAL LAND OFFICE.
June 11th, 1866.

EDWIN BUTLER, ESQ.,
Toulon, Stark Co., Ill.

SIR:

I am in receipt of your letter of the 25th ultimo, making inquiries in regard to the proper mode of subdividing the north tier of sections in Tp. 14 N., R. 6 E., 4th P.M., Ill.

The following remarks, based upon the laws of Congress, touching the subject, will remove your difficulties, and make your case plain:

1st. All corners of the public surveys established by the government surveyors must stand as the true corners they were intended to represent, and the lengths of lines stated in the field notes of the original survey must be considered as the *true lengths* thereof.

2d. Missing corners should be restored to the exact position they originally occupied.

3d. All lines subdividing a section, must be *straight lines* running *through* the section from the corner in one boundary, to its corresponding corner in the opposite boundary of said section.

Now to establish the line between sections 1 and 2 for instance. If the distance from the corner to sections 1, 2, 11, and 12 to the standard line is three chains less than is stated in the field notes, and the $\frac{1}{4}$ section corner cannot be found, it must be established at a *proportionate* distance from each section corner. Thus, suppose the distance stated in the field notes is 40 + 33·50 lks. = 73·50 lks., but by your measurement is only 70·25 lks., it is evident that in order to restore the $\frac{1}{4}$ corner to its original position according to the field notes, your chain must be reduced so that it will make the distances correspond with those of the original survey. This of course will be accomplished by dividing your excess (3·25 lks.) *proportionally* between the distances; therefore,

As 73·50 : 70·25 :: 40 = 38·23 lks.

Also 73·50 : 70·25 :: 33·50 = 32·02 lks.

Hence, by your chaining, you should plant the $\frac{1}{4}$ post 38·23 chs. north of the corner to secs. 1, 2, 11, and 12, and 32·02 south from the township line.

The same principle should govern you in establishing each of the other corners, and all other missing corners of the public surveys, ever keeping in mind that existing original corners must stand, and lost corners must be established at proportionate distances between existing original corners.

When the corners shall have been established, the section will be subdivided by running straight lines from corner to corner.

Very respectfully, etc.,
JAS. M. EDMUNDS, Commissioner.

DEPARTMENT OF THE INTERIOR,
GENERAL LAND OFFICE,
August 30th, 1866.

JAS. C. BRANYAN, ESQ.,
Huntington, Ind.

SIR:

Your letter of the 22d inst., in regard to the proper mode of subdividing Sec. 17, T. 28 N., R. 10 E., Ind., is received. In reply, I have to state in the first place, that the plat of the township on file in this office does not represent sec. 17 as indicated by Fig. 1 in your letter. A copy of sec. 17, taken from our plat, is inclosed herewith.

For your information, however, and to aid you in dealing with similar cases, we will suppose the facts to be as represented by diagrams 1 and 2 in your letter, and base our instructions upon the premises therein stated.

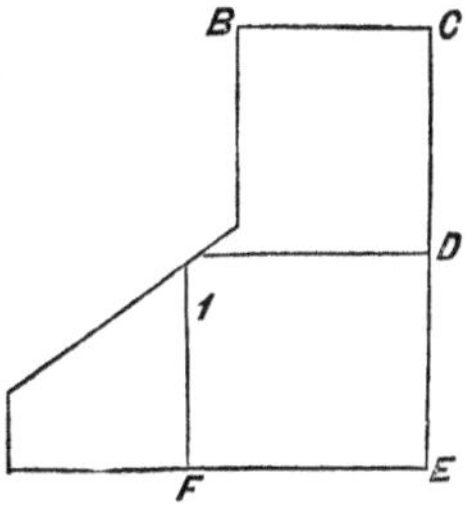

To your first inquiry then, I reply, that the subdivision line between the N. E. and S. E. quarters of the section should be run from the established quarter section corner D, on a due west course, *to its intersection with the Reservation line, be the distance more or less than forty chains.* Then to establish the line between the S. E. and S. W. fractional quarters of the section it is necessary to first ascertain if the *quarter section* post, on the north boundary of the section, was established by the U. S. deputy surveyor; if yes, then the subdivision line between said quarter sections should be a *straight line*, from F projected northward until it intersects the Reservation line, and on a *course* that will intersect the opposite corresponding original quarter section corner.

FIG. 1.

If no such opposite quarter section corner was established by the U. S. deputy surveyor, which is apparently the case in this instance, said corner should be established at 40 chains *proportionate measurement*, west to the section corner C, [by proportionate measurement, is here meant a measurement corresponding with the length of the chain used by the U. S. deputy surveyor, as indicated by the original distances stated in the field notes], and said subdivision line should then be run in the same manner as if the opposite corner had been originally established.

FIG. 2.

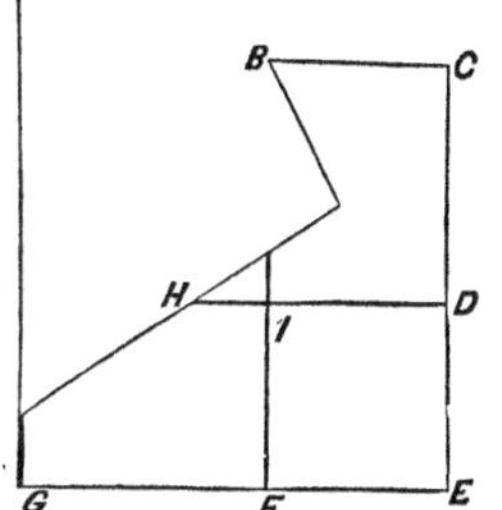

If the distances from the section corner west to the Reservation line should be less than 40 chains, so that no quarter section corner *can be established*, then said subdivision line must be run on a north course from F to its intersection with the boundary of the Reservation; and in either or all of the above cases, the point of intersection of the East and West with the North and South subdivision line run as directed, if outside of the Reservation, will be the true northwest corner of the S. E. quarter section; and the triangular piece of land *H*, if there be any, will of course constitute a fractional northwest quarter.

Very respectfully, etc.,
JOSEPH S. WILSON,
Acting Commissioner.

DEPARTMENT OF THE INTERIOR,
GENERAL LAND OFFICE,
June 14th, 1865.

B. H. TRUSDELL,
Amboy, Ill.

SIR:

I am in receipt of your communication of the 7th inst., inquiring as to the proper mode of subdividing sections into legal subdivisions. In reply, I have to say, from the law of Congress approved Feb. 11th, 1805, the following definite and fixed rules are deduced, to wit:

1st. All corners once established in the field, and approved and returned by the proper officers, shall stand as the true corners which they were intended to designate, even though the intervals do not correspond with the measurements stated in the field notes.

2d. All boundary lines of legal subdivisions which shall not have been actually run and marked in the field, shall be ascertained by running straight lines from the established corners to the opposite corresponding corners.

It will be seen from the foregoing, that the correct mode of dividing sections is by running straight lines from quarter post to opposite quarter post (both North and South and East and West), the common center being determined by the intersection of the lines so run. Great care should be taken in running such subdivision boundaries, to first identify the existing corners as the true original corners established by the U. S. surveyor.

Very respectfully, etc.,
JAS. M. EDMUNDS, Commissioner.

GENERAL LAND OFFICE,
March 30th, 1864.

WARREN BECKWITH, ESQ.,
Geneva, Wis.

SIR:

I am in receipt of your communication of the 23d inst., inquiring as to the proper mode of subdividing sections into legal subdivisions. The law of Congress approved Feb. 11th, 1805 (U. S. Statutes, page 313, Little & Brown's edition), gives explicit directions how this shall be done. This law has not since been repealed or modified, and hence the true and only lawful mode of subdividing sections is the one described therein.

By this law the following definite and fixed rules are enunciated, to wit:

1st. All corners once established in the field, and approved and returned by the proper officers, shall stand as the true corners they were intended to designate, even though the intervals do not correspond with the measurements in the field notes.

2d. All boundary lines of legal subdivisions which shall not have been actually run and marked in the field, shall be ascertained by running straight lines from the established corner to the opposite corresponding corner.

It will be seen from the foregoing rules that the correct mode of dividing sections is by running straight lines from quarter post to opposite quarter posts, the common center being determined by the intersection of the lines so run. Great care should be taken in running such subdivisional boundaries to first identify the existing corners as the true original corners established by the U. S. surveyor.

Very respectfully, etc.,
JOSEPH S. WILSON,
Acting Commissioner.

GENERAL LAND OFFICE,
June 29th, 1863.

D. W. MAXON, ESQ.,
Cedar Creek, Washington Co., Wis.

SIR:

Your letter of the 12th inst., asking for information as to the proper mode of establishing lost corners of the public surveys, etc., is received. As stated in my communication of the 2d inst., this office does not assume to exercise any control over the surveying operations of county surveyors. * * * * * * *

For the information of surveyors who may be called upon to re-establish lost corners of the public surveys or subdivide sections, the following general principles, based upon the laws of Congress and the regulations of the land department in accordance therewith, may be stated:

1st. Section and quarter section corners as established by the government survey, must, by law of Congress, stand as the true corners.

2d. Missing corners must be re-established at the identical point where the original posts were planted by the U. S. deputy surveyors.

3d. The legal presumption is, in the absence of any evidence to the contrary, that lost section and quarter section posts were originally established at the distances indicated in the field notes.

4th. Half quarter section corners must be established equidistant from the section and quarter section posts.

The first proposition above is in accordance with a law of Congress approved February 11th, 1805. To divide a section into quarters a *right line* should be run from the quarter section post in one section line to the corresponding quarter section post in the opposite section line, even though one or more of these posts may have been established nearer to one section corner than the other, thereby giving to one quarter section more than 160 acres and to another less.

The second proposition grows out of the first, and is in accordance with the laws of Congress. It is the duty of the surveyor to re-establish missing posts in the exact locality where they were originally placed in the government survey. The proof of locality first sought to be obtained should be the "witness trees," or any other means of identification contained *in the field notes*, and next, clear and unquestionable testimony of any other kind. If no bearing trees, or other evidences in the field notes or elsewhere exist, by which the locality of the missing posts can be identified or determined in the field, then, as stated under

10

the third head, the legal presumption is, that the missing section or quarter section corners were originally established in conformity with the distances expressed in the field notes, and the surveyor should so re-establish them.

Extinct quarter section corners, except on fractional section lines, when they cannot be identified as above, should be re-established equidistant between the section corners, in a right line between the nearest noted "line trees" each side of it, if there are any, but if none are found, then in a right line between the section corners. Extinct quarter section posts on section lines which close on the north and west boundaries of townships, should be re-established, according to the original measurement thereof, at 40 chains from the last interior section corner.

Extinct section corners may be re-established by running a right line between the nearest noted "line trees" north and south and east and west of the lost corner, if there be any such trees within the distance of the nearest quarter section, or section corners; but if no "line trees" be found, then between the nearest quarter section or section corners, and at the point of intersection of the two lines thus run, establish the section corner, with new bearings, to the nearest and most desirable objects.

The quarter mile posts are not established in government surveys, but are, by law, understood to be equidistant from the section and quarter section corners, as stated under the fourth head, and should be so established by the county surveyor.

It may be remarked, that where the measurement of any section line by the county surveyor does not correspond with the original measurement recorded in the field notes, lost corners should be re-established at *proportional distances* from each other between the known corners.

A proper application of the principles embraced herein will enable the practical surveyor to subdivide the public lands and re-establish the lost corners of the public surveys, in conformity with law and the regulations and usages of the land department.

There are some anomalous cases, such, for instance, as double corners on the north and west boundary lines of townships, an explanation of which must be omitted owing to the length of this communication. The general principles which should govern the county surveyor are, however, indicated with sufficient clearness to guide him in the rightful performance of his duties.

Very respectfully, etc.,

JAS. M. EDMUNDS, Commissioner.

APPENDIX.

APPENDIX.

REGULATIONS IN REGARD TO FURNISHING CERTIFIED COPIES OF PLATS, RECORDS, ETC. ON FILE IN THE GENERAL LAND OFFICE.

APPLICATIONS for certified copies of plats and transcripts and exemplifications of papers, etc., on file in the General Land Office, have become very numerous, and much time is required in answering them. Many of these calls involve the labor of weeks, and in some instances of months, and are of a strictly private character, having no relation to the public interests. When it is proper that such exemplifications should be furnished, they ought manifestly to be furnished at the expense of the parties requiring them. To this end Congress passed the following act, to wit:

"*Be it enacted by the Senate and House of Representatives of the United States of America in Congress assembled*, That, from and after the first day of July next, all exemplifications of patents or papers on file or of record in the General Land Office, which may be required by parties interested, shall be furnished by the Commissioner of said office upon the payment by such parties at the rate of fifteen cents per hundred words, and two dollars for copies of township plats, or diagrams, with an additional sum of one dollar for the Commissioner's certificate of verification with the General Land Office seal; and one of the employees of said office shall be designated by the said Commissioner, as the receiving clerk, and the amounts so received shall, under the direction of the said Commissioner, be paid into the Treasury of the United States; effect to be given to this act according to such regulations as may be prescribed by the Secretary of the Interior not inconsistent with the laws of the United States: *Provided*, That the fees stipulated in the foregoing provisions shall not apply to such authenticated copies as may be required by the officers of any branch of the government, nor to such unverified copies as the Commissioner in his discretion may deem proper to furnish."

Approved July 2d, 1864.

As the law was not approved until the 2d day of July, 1864, it did not go into operation until one year thereafter.

On the 8th day of July, 1865, the Commissioner of the General Land Office, by direction of the Secretary of the Interior, issued the following regulations giving effect to this law:

"*First.*—From and *after July* 1st, 1865, no copies will be furnished until the cost thereof shall first be paid to the General Land Office.

"*Second.*—The applicant must address a communication to the Commissioner of the General Land Office, designating the tract or tracts in regard to which the verified transcripts are wanted, describing as accurately as possible the record, papers, or plats of which transcripts are desired, and sending a sum of money quite sufficient to cover the cost according to the extent of the copying required; and should the sum sent to this office be in excess of the actual cost under the act, such excess will be returned to the applicant.

"The following is the tariff established under the statute for furnishing transcripts, to wit:

"1st. Fifteen (15) cents for every hundred words in a transcript.

"2d. Two dollars ($2) for copy of township plat, or diagram.

"3d. One dollar ($1) for the Commissioner's certificate of verification and official seal.

"4th. One dollar ($1) for appending such certificate and seal to official certificates of approval of assignments of Bounty Land Warrants.

"*Third.*—Upon the receipt at the General Land Office of the application particularly describing the record or papers of which transcripts are required, accompanied by the requisite amount to cover the expense, the same will be duly acknowledged and the exemplifications promptly transmitted."

Applications for certified copies of such documents and papers as can be furnished by the local offices, should be made directly to them, as it is not desirable needlessly to increase this class of work in the General Land Office.

In writing for copies of plats, transcripts, or patents, be particular to give the township, range, and section in a legible hand.

UNCOVERED MEANDERED LAKES.

Small lakes that are not meandered when the surrounding lands are surveyed, are embraced in the legal subdivisions in which they are located respectively and are sold with them. Hence if these lakes "dry up," in whole or in part, the uncovered land belongs with the legal subdivisions in which it is located.

There are other lakes that are meandered and segregated from the public land in the progress of the government surveys, and the contents subtracted from the areas of the surrounding subdivisions. If from any cause such lakes subsequently become dry, the uncovered land belongs to the United States, and any person desiring to purchase such land may have it surveyed under the instructions and conditions on page 88 in regard to unsurveyed islands.

Where a "dried up" meandered lake lies wholly within one legal subdivision, or is properly connected with the section corners, no additional survey is required, but application may be made to the Surveyor-General, requesting that the necessary plats thereof, showing the fractional lots or subdivisions, be prepared and transmitted to the General Land Office and to the Register's office. The affidavits of two respectable persons acquainted with the facts, setting forth that said lake is entirely dry, must accompany the application. In such cases no money will be required to be advanced by the applicants for the plats.

Frequent applications are made to the General Land Office for permission to *drain* certain meandered lakes by artificial means, with the view to acquiring a title to the land so uncovered. *No such permission can be granted under existing laws;* the department deals only with the public *lands.*

ACCRETIONS.—RIPARIAN RIGHTS.

The question of ownership of accretions is a prolific subject of inquiry and correspondence with the General Land Office. Riparian proprietorship is a subject so nearly connected with it, that they may very properly be considered under one head.

It is a principle of the common law that alluvial accretions belong to the coterminous land. Much, however, depends upon the terms and conditions of the grant or conveyance of the original title. The following general rules enunciated by the highest judicial and legal authority in the land will throw some light upon the subject:

"Land gained from the sea, either by alluvian or dereliction, if the same be by little and little, by small and imperceptible degrees, belongs to the owner of the land adjoining."—2 Bl. Com. 261–2.

"The principle governing alluvial accretions gives them to the adjoining owner."—Gerard's Lessee *v.* Hughes et al., 1 Gil and Johnson, 249. "In other words, the description in the original grant gave, in legal effect, to the grantee, a water boundary, and if so, the boundary included the accretions."—18 Howard, 157.

"The rights of riparian proprietors on navigable rivers are limited to high-water mark."—3 Kent Com., 7th ed. 514. "On non-navigable rivers to the thread of the stream."—13 Howard, 397.

"Grants of land bounded by the sea or by navigable rivers where the tide ebbs and flows, extend to high-water mark, that is, to the margin of the periodical flow of the tide, unaffected by extraordinary causes, and the shores below common high-water mark belong to the state in which they are situated. But grants of land bounded on rivers above tidewater, or where the tide does not ebb and flow, carry the grantee to the middle of the river, unless there are expressions in the terms of the grant, or something in the terms taken in connection with the situation and condition of the lands granted that clearly indicate an intention to stop at the edge or margin of the river. There must be a reservation or restriction, express or necessarily implied, which controls the operation of the general presumption and makes the particular grant an exception."

"These are familiar principles of universal application governing the construction of grants of land bounded upon the sea or tide-water, or upon fresh-water rivers, navigable or unnavigable, and whether made by state or individuals, or in large or small tracts."

In the case of public lands, gradual and imperceptible alluvian deposits inure to the coterminous lots, the limits of which are determined by extending the side subdivisional or boundary lines to the water's edge. No additional survey by the government is required in such cases when the lots belong to private parties, but the side lines may be extended by a county surveyor the same as any other lot lines of private lands.

When a river suddenly changes its course, and leaves its original bed, the land so uncovered and lying between the meanders of said stream, inures to the lots which bounded on the river before such change occurred, the thread or center of said uncovered channel constituting the dividing line.

The following case in the Supreme Court of the United States affords a very clear illustration of the proper mode of dealing with accretions. Jones et al. *v.* Johnston, 18 Howard, 150–8. The annexed diagram will help to understand the case.

This suit was brought to recover a portion of alluvion or new-made land in the City of Chicago, formed in Lake Michigan, adjoining the north pier of Chicago harbor, claimed as an increment or accession to lot No. 34.

The plaintiff claimed that a part of its southern termination on the lake was north of the piers and contiguous to the new-formed land, and therefore entitled it to its share of the increment.

The defendant contended that no part of its boundary was on the lake north of the harbor, and therefore no part connected with or adjoining this land new formed.

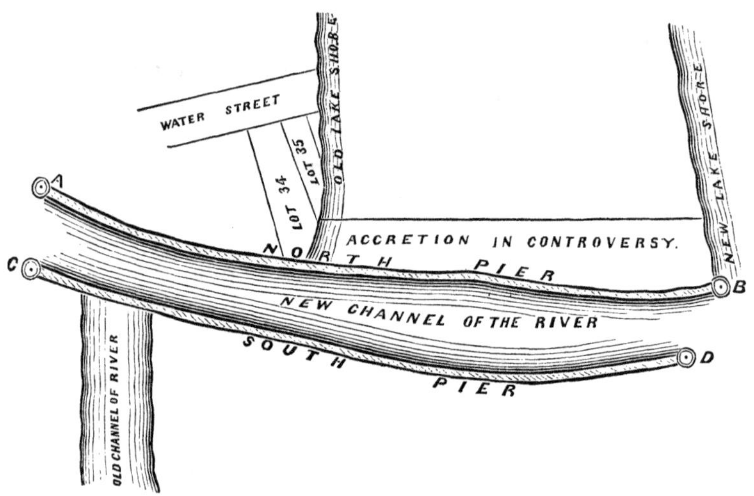

In regard to the point in the case which we are considering, the court held that the inquiry should be made whether or not at the time of the deed to the plaintiff lot No. 34 had a water-line upon the lake north of the north pier of the harbor; if it did, then the question would probably arise in respect to its right to a share of the alluvial accretion formed since that time.

As to the manner of dividing the accretions, the court laid down the rule that each riparian proprietor was entitled to his *proportional share of the entire line* of the newly-made shore.

The case was again before the Supreme Court of the

United States, some six years later, on a bill of exceptions. One point, to which exception was taken, was, "that the court erred in laying down the rule for the partition of the alluvium."

Mr. Justice Swayne answers: "It would be sufficient to say, that the jury having found that lot 34, at the time referred to, had no water front north of the north pier, the question did not arise;" and adds: "But as the views of the court have been misapprehended, and that misapprehension may mislead in other cases, we prefer to deal with the subject as if it were properly before us." He then proceeds to enunciate the rule of the court as follows: "Upon that occasion it was intended to adopt the rule laid down by the Supreme Court of Massachusetts in 17 Pickering, 45, 46, Deerfield *v.* Arms. The court said (Black's Rep. p. 222):

"The rule is—1, to measure the whole extent of the ancient bank or line of the river, and compute how many rods, yards, or feet each riparian proprietor owned on the liver line; 2, the next step is, supposing the former line, for instance, to amount to 200 rods, to divide the newly-formed bank or river line into 200 equal parts, and appropriate to each proprietor as many portions of this *new* as he owned rods on the *old*. When, to complete the division, lines are to be drawn from the points at which the proprietors respectively bounded on the *old*, to the points thus determined, as the points of division on the newly-formed shore. The new lines thus formed, it is obvious, will be either parallel, or divergent, or convergent, according as the *new* shore line of the river equals, or exceeds, or falls short of the old." It is further said: "It may require modification, perhaps, under particular circumstances. For instance, in applying the rule to the ancient margin of the river, to ascertain the extent of such proprietor's title on that margin, the *general* line ought to be taken, and not the actual length of the line on that margin, if it happens to be elongated by deep indentations or sharp projections. In such case, it should be reduced by an equitable and judicious estimate, to the general available line of the land upon the river."

"To this rule we adhere. With the qualification stated, it may be considered as embodying the views of the court upon the subject. In this case, if lot 34 had been found to have had a water front north of the north pier at the time stated, the pier front would have had nothing to do with the partition to be made. The lake front, where the accretion occurred, only could have been regarded. The whole of *that front* should have been taken as the basis of the adjustment."

Strips of land along rivers, bayous, etc. which were omitted when the public surveys were made, are not regarded by the department as coming under the head of riparian proprietorship. It has frequently happened in the prosecution of the public surveys, that deputies have mistaken the banks of bayous or "bottoms" for the true river banks. Instances of this kind occurred on the Kankakee river in Illinois, and on the Missouri river in Iowa. In the latter case the true river bank was found to be more than a mile west of the bank meandered by the deputy surveyor as the margin of the river, and hundreds of acres of public lands have been surveyed and sold between those meanders and the river.

When settlers in any of the public land states, where the Surveyor-General's office has been closed, desire to have such omitted strips of land surveyed, application may be made to the Commissioner of the General Land Office in the same manner and with the same conditions that are required for the survey of small islands.

COAL LANDS.

The act of Congress of 3d March, 1865, *supplementary* to the act of July 1st, 1864, "for the disposal of coal lands and of town property in the public domain," is to enable citizens of the United States who, at the date of the "act, may be in the business of *bona fide* actual coal mining on the public lands, for the purpose of commerce," to enter

160 acres, or less quantity, in legal subdivisions, including their improvements and mining premises, at the minimum price of twenty dollars per acre.

The law, however, expressly *excludes* from its provisions any lands "reserved by the President of the United States for public uses."

The privilege granted is not a general one, but restricted to a single entry by a designated class of individuals, viz.: such as are citizens, and who, on the 3*d March*, 1865, the date of the act, were actually engaged, for "*purposes of commerce*," in "the business of *bona fide* actual coal mining." All persons not so occupied at that date are excluded from the enjoyment of the privilege.

1st. Testimony should be produced satisfactory to the Register and Receiver, showing the party to be a citizen of the United States, and that, at the date of the act, he was engaged "in the business of *bona fide* actual coal mining on the public lands," and "*for the purposes* of commerce."

The facts must be stated in detail, both as to the nature and extent of the coal mining; the period in which the business has been conducted, and in regard to the coal being made by the party an article of commerce, so that a correct judgment may be formed from these facts as to whether the case comes within the purview of the statute. Where the proof is clear and conclusive, the Register and Receiver are authorized to permit the entry, according to "legal subdivisions," in compact form, and so as not to exceed 160 acres.

2d. Where the mining improvements and premises are on land *surveyed* "*at the passage of this act*," it is required that a sworn declaratory statement descriptive of the tract and premises, and also of the extent and character of the improvements, *be filed* within *six months* from the date of the act, and that proof and payment must be made within *one year* from the *date* of such filing.

3d. If the mining premises be on land which may be *surveyed* after the passage of said act, then the declaratory statement shall be filed within three months from the return of the plat to the district land office, and proof and payment must be made within *one year* from the date of such filing.

PUBLIC TIMBER.

The act of Congress, approved May 27th, 1831, makes the cutting or waste or destruction of timber upon the public domain a trespass, and liable to the imposition of a heavy penalty upon conviction in any court of competent jurisdiction. The timber so cut is forfeited, and may be seized and sold for the benefit of the government.

The duty of protecting the public timber has, by order of the Secretary of the Interior, been imposed upon the Registers and Receivers of the respective local land offices.

Pre-emption or homestead settlers are permitted to cut timber for domestic use, buildings, fences, firewood, etc. upon the premises, and for the purpose of clearing and improving the land for cultivation, but are not allowed to cut timber upon the tracts upon which they have settled for the purpose of sale or trade.

TOWN SITES.

The Town Site Law of 1844 was repealed by the act of July 1st, 1864, and the former law is therefore only applicable to cases in which application for entry had been regularly made, and all necessary steps taken prior to July 1st, 1864. There are but few of these old cases remaining undetermined, and it is therefore only necessary in this connection to deal with existing laws in relation to town sites which are:

1st. The act of July 1st, 1864.

2d. The amendatory act of the 3d of March, 1865.

The first provides for founding cities and towns on the public domain, and limits the area of each to 640 acres, and of lots in the same to an area of 4200 square feet.

The amendatory act provides that in towns or cities actually existing at the date of its passage, the lots may be acquired according to existing plats or surveys without restriction as to size.

Actual settlers are entitled to pre-empt one lot and any additional lot on which they may have substantial improvements, at the minimum price, but must prove up and pay for the same, as in ordinary pre-emption cases prior to the day of public sale.

Lots neither claimed by pre-emption nor sold at the public sale will thereafter be subject to private entry by any individuals at the minimum price.

Parties who have already founded or may hereafter found a city or town, are required—

1st. To file with the Recorder of the county in which the town or city is situate a plat thereof, not exceeding 640 acres, describing its exterior boundaries according to the lines of the public surveys, where such surveys have been executed.

2d. Also the plat or map of such city or town must exhibit the name of the city or town, the streets, squares, blocks, lots, and alleys; the size of the same, with measurements and area of each municipal subdivision, the lots in which shall each not exceed 4200 square feet, with a statement of the extent and general character of improvements.

3d. Further, the said map and statement to be verified by oath by the party acting for and in the behalf of the founders of the city or town.

4th. Within one month after filing the map or plat with the Recorder of the county, a verified copy of said map and statement is to be sent to the General Land Office, accompanied by the testimony of two witnesses that such city or town has been established in good faith.

5th. Where the city or town is within the limits of an organized land district, a similar copy of the map and statement must be filed with the Register and Receiver.

6th. The third section provides for cities or towns founded on *unsurveyed* lands, and directs that it may be lawful to adjust the exterior limits of the premises with the lines of the public surveys, where it can be done without impairing the rights of others.

Patents are to issue for all lots sold under the provisions of this act as in ordinary cases.

7th. "By the second section of act of 1st July, 1864, after the transcript and statement have been filed in the General Land Office, the lots are to be offered at public sale to the highest bidder at a *minimum* of ten dollars per lot; but, by the supplemental act, where the *area of each lot* exceeds the *maximum* of 4200 *square feet*, the *minimum price* of each lot shall be increased to such reasonable amount as the Secretary of the Interior may establish," and which price has been established by the Secretary as follows:

"The minimum price of each lot in *a town surveyed before the above-named act of March 3d*, 1865, *took effect*, containing over 4200 square feet and not more than 8400 square feet, shall be fifteen dollars; of each lot containing over 8400 square feet and not more than 12,600 square feet, the minimum price shall be eighteen dollars; of each lot containing over 12,600 square feet and not more than 16,800 square feet, the minimum price shall be twenty dollars; and for larger lots the price shall be increased two dollars for every additional 4200 square feet.

"In the case of out-lots in any such village, town, or city, the minimum price of such out-lots shall be ten dollars; of such out-lots containing more than one acre, the minimum price shall be ten dollars for the first acre, and five dollars for each additional acre in such lot."

TOWN SITE ACT OF 1867.

AN ACT for the relief of the inhabitants of cities and towns upon the public lands.

Be it enacted, etc., That whenever any portion of the public lands of the United States have been or shall be settled upon and occupied as a town site, and therefore not subject to entry under the agricultural pre-emption laws, it shall be lawful in case such town shall be incorporated, for the corporate authorities thereof, and if not incorporated, for the judge of the county court, for the county in which such town may be situated, to enter at the proper land office, and at the minimum price, the land so settled and occupied in trust for the several use and benefit of the occupants thereof, according to their respective interests; the execution of which trust as to the disposal of the lots in such town, and the proceeds of the sale thereof, to be conducted under such rules and regulations as may be prescribed by the legislative authority of the State or Territory in which the same may be situated: *Provided*, That the entry of the land intended by this act to be made shall be made, or a declaratory statement of the purpose of the inhabitants to enter it as a town site under this act, shall be filed with the Register of the proper Land Office prior to the commencement of the public sale of the body of land in which it is included, and that the entry or declaratory statement shall include only such lands as are actually occupied by the town, and the title to which is in the United States. If upon surveyed lands, the entry shall in its exterior limits be made in conformity to the legal subdivisions of the public lands authorized by the act of twenty-fourth of April, one thousand eight hundred and twenty, and when the inhabitants are in number one hundred and less than two hundred shall embrace not exceeding three hundred and twenty acres; and in cases where the inhabitants of such town are more than two hundred and less than one thousand shall embrace not exceeding six hundred and forty acres; and where the number of inhabitants is one thousand thousand shall embrace not exceeding twelve hundred and eighty acres: *Provided*, That for each additional one thousand inhabitants, not exceeding five thousand in all, a further grant of three hundred and twenty acres shall be allowed: *And provided further*, That in any territories in which a land office may not have been established, declaratory statements, as hereinbefore provided, may be filed with the Surveyor-General of the surveying district in which the lands are situate, who shall transmit said declaratory statement to the General Land Office: *And provided further*, That any act of said trustees not made in conformity to the rules and regulations herein alluded to shall be void; effect to be given to the foregoing provisions according to such regulations as may be prescribed by the Secretary of the Interior: *And provided further*, That the provisions of this act shall not apply to military or other reservations heretofore made by the United States, nor to light-houses, custom-houses, mints, or such other public purposes as the interest of the United States may require, whether held under reservations through the Land Office by title derived through the crown of Spain or otherwise: *And provided further*, That no title shall be acquired under the provisions of this act to any mine of gold, silver, cinnabar, or copper.

REPAYMENT OF PURCHASE MONEY AND CHANGES OF ENTRY.

To secure prompt action in this class of cases, the following regulations must be complied with:

1st. The duplicate receipt must be surrendered, and accompany the papers in all cases of application for refunding money submitted by you for the action of this office. Where it has been lost or destroyed, the party applying must advertise it, and give notice of his intention to apply to have refunded the purchase money, which must be inserted weekly, for six weeks, in some paper of extensive circulation in the vicinage of the land. A copy of this must be filed with you, having attached the affidavit of the publisher that it was inserted the requisite number of times.

2d. The applicant must make affidavit that he has not transferred or otherwise encumbered the title to the land. This affidavit may be taken before either of you officially, before a Notary Public using a seal, or a Justice of the Peace; in the last case, a certificate of magistracy must accompany it.

3d. Where a patent has been issued and delivered to the patentee, a deed of relinquishment reconveying the title (conveyed by the patent) to the United States should be made. This deed of relinquishment should be recorded, and a certificate should also be produced from the officer having charge of the books in which conveyances are required to be recorded, showing that said deed is so recorded, and that the records of his office do not exhibit any other conveyance or encumbrance of the title to the land.

4th. In cases of a change of entry, as the purchaser will reasonably wish to hold some evidence of his title until the issuing of the patent, he must first designate the tract he desires to enter, and effect the change of entry, and then across the face of the receipt issued upon the original entry (which will be returned to you from this office when the case has been decided), the Receiver should note the facts of the case, to wit: that the entry has been changed from the tract paid for by said receipt to another, designating the latter, and the number of certificate issued thereon. The receipt must then be delivered to the party, to be surrendered when the patent is delivered to him.

CHANGE OF ENTRY.

The following regulation is prescribed respecting the CHANGES OF ENTRY which may be ordered pursuant to the acts of Congress of 3d March, 1819, and 24th May, 1824, viz.:

"The Receiver will take a receipt from the purchaser for the purchase money of the tract erroneously entered, as if the same was refunded, which receipt will be a voucher to the Receiver's credit, to be introduced into the proper monthly and quarterly accounts," etc.

These acts, it will be observed, do not authorize an actual repayment, which, in fact, it is obvious, was never contemplated by the regulation itself, but merely require the application and transfer of the money which was paid on the wrong entry to the right one; the rule having been adopted as one of official convenience.

By an act of Congress, approved 3d March, 1849, all moneys receivable from customs, and from all other sources, are required to be paid immediately into the Treasury "without abatement or reduction," etc.

To avoid any apparent or real incompatibility in practice with that act, when a change of entry is ordered, pursuant to the acts aforesaid, the Register will issue, as in ordinary cases, his certificate of purchase, only adding to it a MARGINAL NOTE showing the transfer of the payment from the erroneous entry to the new one, referring to the date and number of the former, the date of the authority for the transfer, and making proper references to the proceedings on the books and plats of the local office.

No receipt in such a case is to be issued by the Receiver, unless the area of the new entry is greater than the old one; when, of course, such excess must be paid for, and receipts issued for the same.

The new entries must always be reported in the monthly returns, with an explanatory note showing, in each case, when fully paid for by the transfer of the payment from the old entry, and also when there is a further payment on account of excess; and only the EXCESS PAID FOR is to be carried into the Receiver's QUARTERLY account.

Such changed entries must always appear at the foot of the monthly returns as ADDENDA, and must not affect the aggregate sales, either as to quantity or purchase money, further than the EXCESS in areas, and PURCHASE money for the same.

In any case which may be reported for repayment, and in which the department may decide that the party is legally entitled, such repayment will be made by warrant from the Treasury.

GRADUATED LANDS.

The act of August 4th, 1854, reducing and graduating the prices of the public land to actual settlers thereon, having been repealed by the act of June 2d, 1862, no further entries of land under the said act can be made; and it only remains to close up such entries as have been made and not perfected by the issue of patent. This may be done—

1st. By proving actual settlement, residence, and cultivation, as required by the act of 1854; or

2d. By paying the difference between the graduated price paid at the date of entry, and the regular minimum price of the land at $1.25 per acre (or $2.50, should the tract fall upon a railroad reserved section); or

3d. By abandoning the land to the government, in which case it will become, after the cancellation of the graduated entry, subject to entry under the pre-emption or Homestead Laws; and after public notice of sale, subject to private entry as other public lands.

AGRICULTURAL AND MECHANICAL COLLEGE SCRIP.

This class of land Scrip was authorized by the act of April 2d, 1862, granting aid to agricultural and mechanical colleges in the several States. The extent of the grant is 30,000 acres for each Senator and Representative to which the State is entitled in the Congress of the United States. States which contained a sufficient quantity of public land to satisfy the grant, must take the land within their own boundaries; but those having no United States lands within their limits, may demand Scrip.

Each piece of Scrip represents one quarter section of land—equal to 160 acres.

It can be located only upon a technical quarter section. It will not take a part of two sections, nor a part of two quarters in the same section.

It must be located upon land subject to private entry at $1.25 per acre; that is, land which has been offered at public auction, and not sold, and which remains unoccupied and subject to sale at the price above named.

It cannot be located upon mineral lands, nor can it be received in payment of lands claimed by pre-emption or homestead—nor can more than one milllion of acres be located in any one State.

The Scrip cannot be located by a State, but must be sold and located by the assignees.

There is no restriction as to the quantity which may be located in any Territory.

This Scrip cannot be located upon $2.50 or double minimum lands, although the States in which such lands are situated may select them, accepting *half the quantity* in satisfaction of their claim.

The fee of the local land officers is *four dollars* upon each piece of Scrip.

The States to which Scrip has been issued, or which are entitled to it under the law, are as follows:

	Acres.		Acres.
Connecticut	180,000	Vermont	150,000
Illinois	480,000	Delaware	90,000
Indiana	450,000	Georgia	270,000
Maine	210,000	Kentucky	330,000
Massachusetts	360,000	Maryland	210,000
New Hampshire	150,000	N. Carolina	270,000
New Jersey	210,000	S. Carolina	180,000
New York	990,000	Tennessee	300,000
Ohio	630,000	W. Virginia	150,000
Pennsylvania	780,000	Virginia	300,000
Rhode Island	120,000	Texas	180,000

Total issued and to be issued.......................... 6,990,000

The States not named in the foregoing list select lands within their own limits, and consequently receive no Scrip.

Congress has recently suspended the delivery of Scrip to the States not represented in that body.

INDIAN SCRIP.

This is a class of government land Scrip issued to the Sioux and Chippewa tribes of Indians and to their half-breed descendants, in payment for rights to certain reservations which they relinquished. There is a very limited amount in circulation, and it may be used in the entry of the public lands as follows:

1st. It is not assignable, but must be located in the name and for the use of the Indian or half-breed in whose name it was issued.

2d. The location may be made by the party in person, or by his or her regularly constituted attorney.

3d. The Scrip will lay upon any of the surveyed non-mineral public lands of the United States not reserved or otherwise disposed of.

4th. Upon any unsurveyed public lands upon which the party may have improvements.

5th. Upon any unoccupied lands subject to pre-emption, whether surveyed or unsurveyed.

6th. If filed upon unsurveyed lands, the application must be accompanied by a diagram of the tract, and possession maintained in such a manner as to be fair notice to all other persons of the claim to and occupancy of the premises; and within three months after the survey of the tract by the United States, and the return of the plat of such survey to the local land office, the party must repair to such local office and make claim to the proper legal subdivisions covered by the diagram.

7th. No fees are required to be paid on the location of this class of Scrip, and no receipt is given for the same by the local officers.

8th. Said officers are required to send up to the General Land Office the application, properly certified by themselves, and attach to each application a certificate to the following effect:

"We hereby certify that the within certificate has

day of been located on the con-taining acres agreeably to the section of the article of the treaty of with the and by the party duly authorized to make the location."

Excess receipts are issued for the payment for any fraction less than forty acres or the smallest legal subdivision, which it may be necessary to include in the location, in excess of the number of acres called for by the Scrip.

REVOLUTIONARY BOUNTY LAND SCRIP.

This is a class of Land Scrip issued under the act of the 31st of August, 1852, in favor of the present proprietors of unsatisfied Virginia military land warrants issued or allowed by the authorities of the State of Virginia prior to the 1st of March, 1852.

The Scrip is issued for eighty acre tracts of land, except for fractions, to which the claimant may be entitled, after deducting the eighty acre certificate.

When Scrip is claimed, located or sold by the "guardian of an infant" or the husband of a "femme covert," the evidence of their being such guardian or husband must fully appear.

This Scrip is "assignable by indorsement, attested by two witnesses," in the following manner, upon the back of the certificate:

For *value* received (*I, or we, as the case may be*), the present proprietor of the within certificate of Scrip, do hereby sell and assign the same to of and his heirs and assigns forever.

Witness my hand and seal this the day , 18

Attest: E. F. [SEAL.]

A. B.

C. D.

This Scrip is "receivable in payment of any lands owned by the United States, subject to sale at private entry," except such lands as are claimed by pre-emption, or are settled upon and cultivated, and can be applied at the rate of $1.25 per acre, in the same manner as money, in all cases where the tract applied for contains the area specified in the Scrip, or more; where it contains less, the excess of the Scrip cannot be refunded in money, but may be denoted in the relinquishment as applicable to any other tract.

THE HOMESTEAD LAW.

WHO ARE ENTITLED TO ITS BENEFITS.

Any person who is a citizen of the United States, or has declared his intention to become such under the laws of the United States, and who is:

1st. The head of a family.

2d. Who has arrived at the age of twenty-one years.

3d. Who has served in the army or navy of the United States during actual war, for at least fourteen days, and who has never borne arms against the United States government, or given aid or comfort to its enemies.

Any person possessing the foregoing qualifications may enter 160 acres or a less quantity of the surveyed public lands of the United States; the land to be in a compact body, and in conformity to the lines of the public surveys; unoccupied and subject to entry under the pre-emption laws, at $1.25 per acre. Double minimum or railroad reserved sections may be taken at the double minimum rates, that is, 80 acres of $2.50 land may be taken in lieu of 160 acres at $1.25.

Any person owning and residing on land may enter contiguous land, which, with that already owned and occupied, shall not exceed in the aggregate 160 acres.

The applicant for the benefit of the Homestead Law is required to file with the Register of the United States Local Land Office, for the district in which the lands are located, his application as follows:

(*Form A.*)

HOMESTEAD.

Application. LAND OFFICE at
No. (Date)

I, of do hereby apply to enter, under the provisions of the act of Congress, approved May 20th, 1862, entitled "An act to secure homesteads to actual settlers on the public domain," the of section in township of range containing acres.

[In pre-emption cases the following should be added:

Having filed my pre-emption declaration thereon on day of]

Signature of applicant.

LAND OFFICE at

186 .

I, Register of the Land Office, do certify hereby that the above application is for surveyed lands of the class which the applicant is legally entitled to enter under the Homestead Act of May 20th, 1862, and that there is no prior, valid, adverse right to the same.

Register.

(*Form B.*)

HOMESTEAD.

(*Affidavit.*)

LAND OFFICE at

(Date)

I, of having filed my Application, No , for an entry under the provisions of the act of Congress, approved May 20th, 1862, entitled "An act to secure homesteads to actual settlers on the public domain," do solemnly swear that [*here state whether applicant is the head of a family, or over twenty-one years of age; whether a citizen of the United States, or has filed his declaration of intention of becoming such; or, if under twenty-one years of age, that he has served not less than fourteen days in the army or navy of the United States during actual war; that said Application No. is made for his, or her, exclusive benefit; and that said entry is made for the purpose of actual settlement and cultivation, and not, directly or indirectly, for the use or benefit of any other person or persons whomsoever*].

Sworn to and subscribed, this day of before

[*Register or Receiver*] *of the Land Office.*

(*Form C.*)

HOMESTEAD.

Receiver's Receipt. RECEIVER'S OFFICE

No. (Date)

Application
No.

Received of the sum of dollars cents being the amount of fee, and one-half the compensation of Register and Receiver, for entry of of section in township of range , under the act of Congress, approved May 20th, 1862, entitled "An act to secure homesteads to actual settlers on the public domain."

Receiver.

Should the homestead settler so desire, he may, after settlement and cultivation of the tract entered as above, prove such settlement and cultivation by disinterested witnesses, to the satisfaction of the Commissioner of the General Land Office, when he will have the right to pay for the same, instead of waiting five years for his title as provided in the law; in which case he will receive a certificate in the following form, entitling him to a patent for the land.

LAND OFFICE.
(Date)

CERTIFICATE. No. APPLICATION. No.

IT IS HEREBY CERTIFIED, That, pursuant to the provisions of the act of Congress approved May 20th, 1862, entitled "An act to secure homesteads to actual settlers on the public domain," has made payment in full for of section in township of range , containing acres.

NOW, THEREFORE, BE IT KNOWN, That on presentation of this certificate to the Commissioner of the General Land Office, the said shall be entitled to a patent for the tract of land above described.

Register.

The fee required to be paid on a homestead entry is $10 for the use of the government, and one per cent. *commissions* to each of the officers—Register and Receiver—on the value of the land, at $1.25 or $2.50 per acre, as the case may be. A like commission has also to be paid upon receiving a Patent Certificate.

Lands entered under this act are not liable for debts contracted prior to the issue of patent.

An abandonment of the land for more than six months within the five years residence required, is cause of forfeiture, unless the land shall have been paid for under the 8th section of the act as above explained.

In case of death of the homestead settler, the land may be sold for the benefit of infant heirs, but for no other purpose.

Pre-emptions upon surveyed $1.25 lands may be transmuted into homesteads if no other rights shall have intervened.

Persons residing at a distance from the proper land office may make application before the clerk of the county in which he or she resides, and transmit the same with the necessary fee and commissions to the Register and Receiver, in which case the same forms will be used, only making the necessary alterations as to the officer before whom the affidavit and application are made.

Persons actually engaged in the military or naval service of the United States, may enter land under the Homestead Act, by duly authorized agents or attorneys. The necessary affidavits and applications, as well as the power of attorney must be verified before the immediate commanding officer. The time which he remains in the service will be counted as residence upon and occupancy of the land; but, upon leaving the service, he must within a reasonable time repair to the land, and remain upon, and cultivate the same in good faith, or his claim will become forfeited.

The Homestead Law has been further modified by the act of Congress, approved June 21st, 1866, which is fully explained in the following circular and copy of the statute:

DEPARTMENT OF THE INTERIOR,
GENERAL LAND OFFICE,
September 25th, 1866.

GENTLEMEN:

Annexed is the act of Congress "approved June 21st, 1866," providing for the disposal of the public lands for homestead actual settlement in the States of Alabama, Mississippi, Louisiana, Arkansas, and Florida.

The 1st section of this act, in providing for the disposal of the public lands in the States above mentioned, according to the provisions of the Homestead Act of May 20th, 1862, and the amendatory act of March 21st, 1864, restricts entries to not more than a half-quarter section or 80 acres when held at $1.25 per acre. Should the tract selected, however, be $2.50 per acre land, only half that quantity, or 40 acres, can be entered according to the principle fixed in the original statute of 1862, the law imposing this restriction as to quantity in said States for two years from its passage. After the expiration of that time, however, entries as to quantity of acres, should no other legislation be had, will be governed by the provisions of the 1st section of the act of the 20th May, 1862

This section of the act of 1866 provides further that the benefits of the law shall be extended to citizens of the United States without distinction or discrimination as to race or color, and that no *mineral lands* shall be liable to homestead entry for settlement under its provisions. In lieu of the $10 fee required by act of May 20th, 1862, to be paid at the time of entry, the sum of $5 is to be paid *at the time of the issue of patent* in each case. The commissions of the Register and Receiver will be the same as provided for in acts of May 20th, 1862, and March 21st, 1864, viz., one per cent. on the cash value of the land to each officer, with additional 50 per centum provided for in the 6th section of the act of March 21st, 1864, to the officers in the several regions of country therein named.

The foregoing provisions are mainly of *special* application to the States first above mentioned. The 2d section of the act is of *general* application to *all* the States and Territories, and re-enacts the 2d section of the act of May 20th, 1862, with the following modifications, viz.:

Until the 1st of January, 1867, the applicant is required to make affidavit, in addition to the oath required by said section, that he has not borne arms against the United States or given aid and comfort to its enemies. This requirement is already inserted in the form B of affidavit adopted in the administration of the original act, and forms part of the instructions in circular of October 30th, 1862. The effect, then, of this legal stipulation is to limit that particular requirement to the 1st of January, 1867.

The law in question is further of general application in this, that the fee is reduced to $5 when the entry shall not embrace more than *eighty* acres held at $1.25 per acre; but where the entry is in excess of that quantity, the usual fee ($10) must be paid. The 3d section of the act stipulates that all the provisions of the Homestead Law of May 20th, 1862, and the amendatory act of 1864, so far as the same may be applicable, except as modified by act of June 21st, 1866, are to be held and regarded a part of said act of 1866 as fully as if therein enacted and set forth.

Very respectfully, your obedient servant,

JOS. S. WILSON, Commissioner.

TO THE REGISTERS AND RECEIVERS
of the United States Land Offices.

AN ACT for the disposal of the public lands for homestead actual settlement in the States of Alabama, Mississippi, Louisiana, Arkansas, and Florida.

Be it enacted by the Senate and House of Representatives of the United States of America in Congress assembled, That, from and after the passage of this act, all the public lands in the States of Alabama, Mississippi, Louisiana, Arkansas, and Florida shall be disposed of according to the stipulations of the homestead law of twentieth May, eighteen hundred and sixty-two, entitled "An act to secure homesteads to actual settlers on the public domain," and the act supplemental thereto, approved twenty-first of March, eighteen hundred and sixty-four, but with this restriction, that until the expiration of two years from and after the passage of this act, no entry shall be made for more than a half-quarter section, or eighty acres; and in lieu of the sum of ten dollars required to be paid by the second section of said act, there shall be paid the sum of five dollars at the time of the issue of each patent; and that the public lands in said States shall be disposed of in no other manner after the passage of this act: *Provided*, That no distinction or discrimination shall be made in the construction or execution of this act on account of race or color: *And provided further*, That no mineral lands shall be liable to entry and settlement under its provisions.

SEC. 2. *And be it further enacted*, That section second of the above-cited homestead law, entitled "An act to secure homesteads to actual settlers on the public domain," approved May twentieth, eighteen hundred and sixty-two, be so amended as to read as follows: That the person applying for the benefit of this act shall, upon application to the Register of the land office in which he or she is about to make such entry, make affidavit before the said Register or Receiver that he or she is the head of a family, or is twenty-one years or more of age, or shall have performed service in the army or navy of the United States, and that such application is made for his or her exclusive use and benefit, and that said entry is made for the purpose of actual settlement and cultivation, and not either directly, or indirectly, for the use or benefit of any other person or persons whomsoever; and upon filing the said affidavit with the Register or Receiver, and on payment of five dollars, when the entry is not more than eighty acres, he or she shall thereupon be permitted to enter the amount of land specified: *Provided, however*, That no certificate shall be given or patent issued therefor until the expiration of five years from the date of such entry; and if, at the expiration of such time, or at any time within two years thereafter, the person making such entry, or, if he be dead, his widow, or, in case of her death, his heirs or devisee, or, in case of a widow making such entry, her heirs or devisee, in case of her death, shall prove by two credible witnesses that he, she, or they have resided upon or cultivated the same for the term of five years immediately succeeding the time of filing the affidavit aforesaid, and shall make affidavit that no part of said land has been alienated, and that he will bear true allegiance to the government of the United States; then, in such case, he, she, or they, if at that time a citizen of the United States, shall be entitled to a patent, as in other cases provided by law: *And provided further*, That in case of the death of both father and mother, leaving an infant child or children under twenty-one years of age, the right and fee shall inure to the benefit of said infant child or children; and the executor, administrator, or guardian may, at any time within two years after the death of the surviving parent, and in accordance with the laws of the State in which such children, for the time being, have their domicile, sell said land for the benefit of said infants, but for no other purpose, and the purchaser shall acquire the absolute title by the purchase, and be entitled to a patent from the United States on the payment of the office fees and sum of money herein specified: *Provided*, That until the first day of January, eighteen hundred and sixty-seven, any person applying for the benefit of this act shall, in addition to the oath hereinbefore required, also make oath that he has not borne arms against the United States, or given aid and comfort to its enemies.

SEC. 3. *And be it further enacted,* That all the provisions of the said homestead law, and the act amendatory thereof, approved March twenty-first, eighteen hundred and sixty-four, so far as the same may be applicable, except so far as the same are modified by the preceding sections of this act, are applied to and made part of this act as fully as if herein enacted and set forth.

Approved June 21st, 1866.

By the terms of the Amendatory Act of June 21st, 1866, the restriction which prohibits persons who have borne arms against the United States or given aid or comfort to the enemies thereof, ceased to be operative on the 1st of January, 1867, and any citizen of the United States possessing the other qualifications requisite may now enter land under the Homestead Law.

Mineral lands in the States named in the act of 21st June, 1866, do not fall under the restrictions of that act, and may therefore be entered under the general laws.

The Homestead Law being now of general application over the whole of the surveyed public lands, and the law under which a large majority of actual settlers will acquire titles to their homes, it is of the utmost importance that its provisions should be fully understood. For this reason the law with the amendments of 1864 and 1866 are here given.

AN ACT to secure homesteads to actual settlers on the public domain.

Be it enacted by the Senate and House of Representatives of the United States of America in Congress assembled, That any person who is the head of a family, or who has arrived at the age of twenty-one years, and is a citizen of the United States, or who shall have filed his declaration of intention to become such, as required by the naturalization laws of the United States, and who has never borne arms against the United States government or given aid and comfort to its enemies, shall, from and after the first January, eighteen hundred and sixty-three, be entitled to enter one quarter section or a less quantity of unappropriated public lands, upon which said person may have filed a pre-emption claim, or which may, at the time the application is made, be subject to pre emption at one dollar and twenty-five cents, or less, per acre; or eighty acres or less of such unappropriated lands, at two dollars and fifty cents per acre, to be located in a body, in conformity to the legal subdivisions of the public lands, and after the same shall have been surveyed: *Provided,* That any person owning and residing on land may, under the provisions of this act, enter other land lying contiguous to his or her said land, which shall not, with the land so already owned and occupied, exceed in the aggregate one hundred and sixty acres.

SEC. 2. *And be it further enacted*, That the person applying for the benefit of this act shall, upon application to the register of the land office in which he or she is about to make such entry, make affidavit before the said register or receiver that he or she is the head of a family, or is twenty-one or more years of age, or shall have performed service in the army or navy of the United States, and that he has never borne arms against the government of the United States or given aid and comfort to its enemies, and that such application is made for his or her exclusive use and benefit, and that said entry is made for the purpose of actual settlement and cultivation, and not, either directly or indirectly, for the use or benefit of any other person or persons whomsoever; and upon filing the said affidavit with the register or receiver, and on payment of ten dollars, he or she shall thereupon be permitted to enter the quantity of land specified: *Provided, however*, That no certificate shall be given or patent issued therefor until the expiration of five years from the date of such entry: and if, at the expiration of such time, or at any time within two years thereafter, the person making such entry—or if he be dead, his widow: or in case of her death, his heirs or devisee: or in case of a widow making such entry, her heirs or devisee, in case of her death—shall prove by two credible witnesses that he, she, or they have resided upon or cultivated the same for the term of five years immediately succeeding the time of filing the affidavit aforesaid, and shall make affidavit that no part of said land has been alienated, and that he has borne true allegiance to the government of the United States; then, in such case, he, she, or they, if at that time a citizen of the United States, shall be entitled to a patent, as in other cases provided for by law: *And provided, further*, That in case of the death of both father and mother, leaving an infant child or children under twenty-one years of age, the right and fee shall inure to the benefit of said infant child or children; and the executor, administrator, or guardian may, at any time within two years after the death of the surviving parent, and in accordance with the laws of the State in which such children for the time being have their domicile, sell said land for the benefit of said infants, but for no other purpose: and the purchaser shall acquire the absolute title by the purchase, and be entitled to a patent from the United States, on payment of the office fees and sum of money herein specified.

SEC. 3. *And be it further enacted*, That the register of the land office shall note all such applications on the tract books and plats of his office, and keep a register of all such entries, and make return thereof to the General Land Office, together with the proof upon which they have been founded.

SEC. 4. *And be it further enacted*, That no lands acquired under the provisions of this act shall in any event become liable to the satisfaction of any debt or debts contracted prior to the issuing of the patent therefor.

SEC. 5. *And be it further enacted*, That, if at any time after the filing of the affidavit, as required in the second section of this act and before the expiration of the five years aforesaid, it shall be proven, after due notice to the settler, to the satisfaction of the register of the land office, that the person having filed such affidavit shall have actually changed his or her residence, or abandoned the said land for more than six months at any time, then and in that event the land so entered shall revert to the government.

SEC. 6. *And be it further enacted,* That no individual shall be permitted to acquire title to more than one quarter section under the provisions of this act; and that the Commissioner of the General Land Office is hereby required to prepare and issue such rules and regulations, consistent with this act, as shall be necessary and proper to carry its provisions into effect, and that the registers and receivers of the several land offices shall be entitled to receive the same compensation for any lands entered under the provisions of this act that they are now entitled to receive when the same quantity of land is entered with money, one-half to be paid by the person making the application at the time of so doing, and the other half on the issue of the certificate by the person to whom it may be issued; but this shall not be construed to enlarge the maximum of compensation now prescribed by law for any register or receiver: *Provided,* That nothing contained in this act shall be so construed as to impair or interfere in any manner whatever with existing pre-emption rights: *And provided further,* That all persons who may have filed their applications for a pre-emption right prior to the passage of this act shall be entitled to all privileges of this act: *Provided further,* That no person who has served, or may hereafter serve, for a period of not less than fourteen days in the army or navy of the United States, either regular or volunteer, under the laws thereof, during the existence of an actual war, domestic or foreign, shall be deprived of the benefits of this act on account of not having attained the age of twenty-one years.

SEC. 7. *And be it further enacted,* That the fifth section of the act entitled "An act in addition to an act more effectually to provide for the punishment of certain crimes against the United States, and for other purposes," approved the third of March, in the year eighteen hundred and fifty-seven, shall extend to all oaths, affirmations, and affidavits required or authorized by this act.

SEC. 8. *And be it further enacted,* That nothing in this act shall be so construed as to prevent any person who has availed him or herself of the benefits of the first section of this act from paying the minimum price, or the price to which the same may have graduated, for the quantity of land so entered at any time before the expiration of the five years, and obtaining a patent therefor from the government, as in other cases provided by law, on making proof of settlement and cultivation as provided by existing laws granting pre-emption right.

Approved May 20th, 1862.

AN ACT amendatory of the Homestead Law, and for other purposes.

Be it enacted by the Senate and House of Representatives of the United States of America in Congress assembled, That in case of any person desirous of availing himself of the benefits of the Homestead Act of twentieth May, eighteen hundred and sixty-two, but who, by reason of actual service in the military or naval service of the United States, is unable to do the personal preliminary acts at the district land office which the said act of twentieth May, eighteen hundred and sixty-two, requires, and whose family, or some member thereof, is residing on the land which he desires to enter, and upon which a bona fide improvement and settlement have been made, it shall and may be lawful for such person to make the affidavit required by said act before the officer commanding in the branch of the service in which the party may be engaged, which affidavit shall be as binding in law, and with like penalties, as if taken before the Register or Receiver; and upon such affidavit being filed with the Register by the wife or other representative of the party, the same shall become effective from the date of such filing, provided the said application and affidavit are accompanied by the fee and commissions as required by law.

SEC. 2. *And be it further enacted*, That, besides the ten-dollar fee exacted by the said act, the homestead applicant shall hereafter pay to the Register and Receiver each, as commissions, at the time of entry, one per centum upon the cash price as fixed by law of the land applied for, and like commissions when the claim is finally established and the certificate therefor issued as the basis of a patent.

SEC. 3. *And be it further enacted*, That in any case hereafter in which the applicant for the benefit of the homestead, and whose family, or some member thereof, is residing on the land which he desires to enter, and upon which a bona fide improvement and settlement have been made, is prevented, by reason of distance, bodily infirmity, or other good cause, from personal attendance at the district land office, it shall and may be lawful for him to make the affidavit required by the original statute before the clerk of the court for the county in which the applicant is an actual resident, and to transmit the same, with the fee and commissions, to the Register and Receiver.

SEC. 4. *And be it further enacted*, That in lieu of the fee allowed by the twelfth section of the pre-emption act of fourth September, eighteen hundred and forty-one, the Register and Receiver shall each be entitled to one dollar for their services in acting upon pre-emption claims, and shall be allowed, jointly, at the rate of fifteen cents per hundred words, for the testimony which may be reduced by them to writing for claimants, in establishing pre-emption or homestead rights; the regulations for giving proper effect to the provisions of this act to be prescribed by the Commissioner of the General Land Office.

SEC. 5. *And be it further enacted*, That where a pre-emptor has taken the initiatory steps required by existing laws in regard to actual settlement, and is called away from such settlement by being actually engaged in the military or naval service of the United States, and by reason of such absence is unable to appear at the district land office to make, before the Register or Receiver, the affidavits required by the thirteenth section of the pre-emption act of fourth September, eighteen hundred and forty-one, the time for filing such affidavit and making final proof and entry of location shall be extended six months after the expiration of his term of service, upon satisfactory proof, by affidavit or the testimony of witnesses, that the said pre-emptor is so in the service, being filed with the Register of the Land Office for the district in which his settlement is made.

SEC. 6. *And be it further enacted*, That the Registers and Receivers in the State of California, in the State of Oregon, and in the Territories of Washington, Nevada, Colorado, Idaho, New Mexico, and Arizona, shall be entitled to collect and receive, in addition to the fees and allowances provided by this act, fifty per centum of said fees and allowances as compensation for their services: *Provided*, That the salary and fees allowed any Register and Receiver shall not exceed in the aggregate the sum of three thousand dollars per annum.

Approved March 21st, 1864.

MILITARY BOUNTY LAND WARRANTS.

This class of land warrants has been issued under various acts of Congress; but most of the warrants in circulation are authorized by the acts 1847, 1850, 1852, or 1855, all of which by subsequent legislation have been made to assimilate in character, and are now locatable under the same general rules and regulations of the department, which are as follows:

1st. These warrants are now assignable, and properly transferred, carry all the rights of the original warrantee. To make the transfer valid the assignee must be named, as the department will not recognize assignments in blank. The assignment must in all cases be verified by two witnesses. Assignments by Indians must be executed before the United States Indian Agent for the tribe of which the assignee is a member. If not a member of any tribe, the Indian may assign as other parties.

Warrants cannot be located upon occupied land, except it be subject to any legal right to which such occupant may prove to be entitled.

ASSIGNMENT OF LAND WARRANTS AND LOCATIONS.

By the first section of the act of Congress entitled "An act making land warrants assignable, and for other purposes," approved March 22d, 1852, it is provided: "That all warrants for military bounty land which have been, or may hereafter be issued, under any law of the United States, and all valid locations of the same, which have been, or may hereafter be made, are hereby declared to be assignable, by deed or instrument of writing, made and executed after the taking effect of this act, according to such form and pursuant to such regulations as may be prescribed by the Commissioner of the General Land Office, so as to vest the assignee with all the rights of the original owners of the warrant or location."

In accordance with the provisions of this section, the following forms are prescribed by said commissioner for the assignment of the warrants and locations referred to, to wit:

FORM FOR THE ASSIGNMENT OF THE WARRANT—NO. 1.

For value received, I, A. B., to whom the within warrant, No.   , was issued, do hereby sell and assign unto C. D., of   and to his heirs and assigns forever, the said warrant, and authorize him to locate the same, and receive a patent therefor.

Witness my hand and seal this day of , 18 .

Attest—(*Two witnesses.*) A. B. [SEAL.]

FORM OF ACKNOWLEDGMENT WHERE THE VENDOR IS KNOWN TO THE OFFICER TAKING THE ACKNOWLEDGMENT.

State of , County of

On this day of in the year , before me, personally came (*here insert the name of the warrantee*) to me well known, and acknowledged the foregoing assignment to be his act and deed; and I certify that the said (*here insert the name of the warrantee*) is the identical person, to whom the within warrant issued, and who executed the foregoing assignment thereof.

(*Officer's signature.*)

FORM OF ACKNOWLEDGMENT WHERE THE VENDOR IS NOT KNOWN TO THE OFFICER, AND HIS IDENTITY HAS TO BE PROVED.

State of , County of

On this day of , in the year , before me, personally came (*here insert the name of the warrantee*) and (*here insert the name and residence of a witness*) being well known to me as a credible and disinterested person, was duly sworn by me, and on his oath declared and said, that he well knows the said (*here insert the name of the warrantee*) and that he is the same person to whom the within warrant issued, and who executed the foregoing assignment, and his testimony being satisfactory evidence to me of that fact, the said (*here insert the name of the warrantee*) thereupon acknowledged the said assignment to be his act and deed.

(*Officer's signature.*)

FORM FOR THE ASSIGNMENT OF THE LOCATION—NO. 2.

For value received, I, A. B., to whom the within certificate of location was issued, do hereby sell and assign unto C. D., and to his heirs and assigns forever, the said certificate of location, and the warrant and land therein described, and authorize him to receive his patent therefor.

Witness my hand and seal this day of , 18 .

Attest—(*Two witnesses.*) A. B. [SEAL.]

FORM OF ACKNOWLEDGMENT WHERE THE VENDOR IS PERSONALLY KNOWN TO THE OFFICER TAKING THE SAME.

State of , County of

On this day of in the year , before me, personally came (*here insert the name of the person to whom the certificate of location issued*) to me well known, and acknowledged the foregoing assignment to be his act and deed; and I certify that the said (*here insert the name of the person to whom the certificate of location issued*) is the identical person to whom the within certificate of location issued, and who executed the foregoing assignment thereof.

(*Officer's signature.*)

FORM OF ACKNOWLEDGMENT WHERE THE VENDOR IS NOT PERSONALLY KNOWN TO THE OFFICER, AND WHERE HIS IDENTITY HAS TO BE PROVED.

State of , County of

On this day of , in the year , before me, personally came (*here insert the name of the person to whom the certificate of location issued*) and (*here insert the name and residence of a witness*) being well known to me as a credible and disinterested person, was duly sworn by me, and on his oath declared and said, that he well knows the said (*here insert the name of the person to whom the certificate of location issued*) and that he is the same person to whom the within certificate of location issued, and who executed the foregoing assignment; and his testimony being satisfactory evidence to me of that fact, the said (*here insert the name of the person to whom the certificate of location issued*) thereupon acknowledged the said assignment to be his act and deed.

(*Officer's signature.*)

Assignment No. 1 and acknowledgment must be indorsed upon the warrant, and No. 2 and acknowledgment upon the certificate of location; and must be attested by two witnesses acknowledged before a Register or Receiver of a Land Office, a Judge of a Court of Record, a Justice of the Peace, or a Commissioner of Deeds, resident in the State from which he derives his appointment; and in every instance where the acknowledgment is made before either of the officers above specified, except the *Register or Receiver* of a Land Office, it must be accompanied by a certificate, under seal of the proper authority, of the official character of the person before whom the acknowledgment was made, and also of the genuineness of his signature.

All assignments of bounty land warrants issued under the act of September 28th, 1850, made before the date of this act, are invalid and void.

By these provisions, where the lands are subject to private entry at $1.25 per acre, the holder of an eighty-acre warrant can take any two forty-acre lots, forming a compact body of eighty acres; and the holder of a warrant for one hundred and sixty acres can take two eighty-acre or four forty-acre tracts, forming a compact body of one hundred and sixty acres.

This act does not authorize the holder of an eighty-acre warrant to locate therewith a forty-acre tract of land at $2.50 per acre in full satisfaction thereof, but he must locate, by legal subdivisions, the compact body of eighty acres as near as may be, and pay the difference in cash. So also of one hundred and sixty-acre warrants.

Where parties may desire to avail themselves of the privilege of having their warrants located through the General Land Office, as provided for by the act of 28th September, 1850, they must take the necessary steps to pay to the Register and Receiver the fees to which they are entitled. The same course must be observed by persons remote from the district land officers in making applications by letter to those officers. Without the payment of those fees the warrants cannot be located.

By the terms of this law, the fees of officers are as follows:

For a 40-acre warrant, fifty cents each to Register and Receiver. Total $1.

For an 80-acre warrant, one dollar each to Register and Receiver. Total $2.

For a 160 acre warrant, two dollars each to Register and Receiver. Total $4.

The following is a form giving authority to sell warrants and locations under powers of attorney, which, however, must invariably be indorsed on the warrant, or they will not be recognized:

FORM OF A POWER OF ATTORNEY.

Know all men by these presents, that I (*here insert the name of the warrantee*), of the County of and State of do hereby constitute and appoint of my true and lawful attorney, for me, and in my name, to sell and convey the within land warrant, No. for acres of land, which issued under the act of September, 1850.

Signed in presence of

(*Warrantee's signature.*)

The acknowledgment of this power of attorney must be taken and certified in the same manner as the acknowledgments of the sales of the warrant or certificate of location hereinbefore prescribed, and must also be indorsed on the warrant.

GENERAL LAND OFFICE,
November 1st, 1858.

Annexed is a copy of the act of Congress, approved June 3d, 1858, "declaring the title to land warrants in certain cases."

In virtue of this act, the title to land warrants issued *after* the death of the warrantee vests in the "widow," if there be one, and if there be none, in the heirs or legatees of the claimant.

All such warrants are declared "personal chattels," and may be assigned by "such widows," "heirs," or "legatees," or by the legal representatives of the deceased claimant, for the use of such heirs or legatees only.

To make a warrant, when issued to deceased persons, available, it should be accompanied by a certificate, under seal, from a court having jurisdiction of probate matters, giving the *date* of the *decease* of the claimant.

1st. If he died *before* the date of the warrant, then the name of his "*widow*" should be stated in that certificate, if there be one, whose assignment will be sufficient in the ordinary form.

2d. If no widow, that fact should appear in the certificate of the probate court showing the names of the heirs, and only heirs-at-law, of the claimant, naming such as are *adults*, and such as are *minors*. If all are adults, then their simple transfer is all that is required on the warrant, to which the certificate of the probate court must be appended; if some, or all, are minors, they may assign by their guardians, whose letter of guardianship should also be appended.

3d If the claimant died *after* the date of the warrant, then the title thereto descends according to the law of domicile.

In this class of cases, if the claimant died *intestate*, there should be a certificate from the probate court giving the names of the heirs, and only heirs-at-law, who, if adults, may assign, as in ordinary cases; and if minors, may assign by guardians, as aforesaid.

If the warrantee died *testate*, a certified transcript of the will should be annexed, with an assignment by the legatees or by the executors, where the will does not specifically dispose of the warrant; but in that case a transcript of the letters testamentary must accompany the transfer.

4th. Or, in any of the foregoing cases of *intestacy*, the warrant may be assigned by the administrator of the decedent, as his legal representative, "for the use of the heirs only;" but the assignment must be accompanied by a certified copy of the letters of administration.

5th. In virtue of the 2d section of said act of 3d June, 1858, warrants under the act of 1855, as well as those issued under previous laws, may be applied to lands "which are subject to entry at a greater minimum than" $1.25 per acre, by the locator paying "in cash, the difference between the value of such warrants at one dollar and twenty-five cents per acre, and the tract of land located on."

Thus, for example, 160-acre warrant may be located on 160 acres at $2.50 per acre, and the difference, $200, paid in cash, or

2 warrants of 80 acres each, or 4 warrants of 40 acres, may be applied to 160-acre tract—each, however, to be located on a specific legal subdivision of the 160 acres—and the difference, $200, must in all cases be paid in cash.

6th. In regard to all pre-emptions, at *one dollar and twenty-five cents* per acre, it is held, that a pre-emptor may use one, two, or more warrants in locating the land pre-empted, each warrant to cover a specific subdivision of the land—that is, a 40-acre warrant must be located on a specific 40-acre tract, an 80 on an 80-acre tract, and so on.

When a subdivision is fractional, and overruns the number of acres called for by the warrant, the fractional excess must be paid for in cash.

The following general sections are added for the information of parties interested:

7th. Patents for bounty land locations are issued in the exact order of date of location; and are sent to the district land office for delivery, unless, before the transmission, a party files, in this office, the duplicate certificate, when the patent will be sent to such address as the owner may indicate.

8th. When the duplicate certificate is lost, the patent will be delivered, upon the patentee filing in this office his affidavit, stating the fact of its loss, and that it was not assigned by him, and that he is the present *bona fide* owner of the land.

9th. When an original warrant which had been assigned is lost, and a duplicate warrant is issued in lieu of it, a new assignment must be indorsed thereon from the warrantee, or in default of that, a decree of title must be obtained from a court of competent jurisdiction, and a transcript thereof appended to the duplicate warrant.

AN ACT declaring the title to land warrants in certain cases.

Be it enacted by the Senate and House of Representatives of the United States of America in Congress assembled, That when proof has been or shall hereafter be filed in the Pension Office, during the lifetime of a claimant, establishing, to the satisfaction of that office, his or her right to a warrant for military services, and such warrant has not been or may not hereafter be issued until after the death of the claimant, and all such warrants as have been heretofore issued subsequent to the death of the claimant, the title to such warrants shall vest in the widow, if there be one, and if there be no widow, then in the heirs or legatees of the claimant; and all such warrants, and all other warrants issued pursuant to existing laws, shall be treated as personal chattels, and may be conveyed by assignment of such widow, heirs, or legatees, or by the legal representatives of the deceased claimant, for the use of such heirs or legatees only.

Sec. 2. *And be it further enacted,* That the provisions of the first section of the act approved March twenty-two, eighteen hundred and fifty-two, to make land warrants assignable, and for other purposes, shall be so extended as to embrace land warrants issued under the act of the third March, eighteen hundred and fifty-five.

Approved June 3d, 1858.

AN ACT to authorize the reissue of land warrants in certain cases, and for other purposes.

Be it enacted by the Senate and House of Representatives of the United States of America in Congress assembled, That whenever it shall appear that any certificate or warrant, issued in pursuance of any law of the United States granting bounty land, has been lost or destroyed, whether the same had been sold and assigned by the warrantee or not, the Secretary of the Interior shall be, and he is hereby, authorized and required to cause a new certificate or warrant of like tenor to be issued in lieu thereof; which new certificate or warrant may be assigned, located, and patented in like manner as other certificates or warrants for bounty land are now authorized by law to be assigned, located, and patented; and in all cases where warrants have been or may be reissued, the original warrant, in whosever hands it may be, shall be deemed and held to be null and void, and the assignment thereof, if any there be, fraudulent; and nò patent shall ever issue for any land located therewith, unless such presumption of fraud in the assignment be removed by due proof that the same was executed by the warrantee in good faith and for a valuable consideration.

SEC. 2. *And be it further enacted*, That the said Secretary of the Interior shall be, and he is hereby, authorized and required to prescribe such rules and regulations for carrying this act into effect as he may deem necessary and proper in order to protect the government against imposition and fraud by persons claiming the benefit of this act; and all laws and parts of laws for the punishment of false swearing and frauds against the United States are hereby made applicable to false swearing and fraud under this act.

Approved June 23d, 1860.

1st. Whenever a warrant has failed to reach the hands of the party entitled to receive it, and to whom it was sent, or has been lost or destroyed after having been received, in order to prevent the issuing of a patent to a fraudulent holder of the same, the actual owner must at once file in the General Land Office a caveat in the form of an affidavit, duly authenticated, setting forth the nature of his title to the warrant, and the particulars as to its loss, and giving his post-office address.

2d. He must give public notice of the facts in the case, at least once a week for six successive weeks, in some newspaper of general circulation published at or nearest the place to which the warrant was directed, or where the loss occurred. In such publication (a copy of which must be furnished to Commissioner of

Pensions, with the affidavit of the publisher as to its due appearance) the intention must also be expressed of applying to the Commissioner of Pensions for a reissue of the lost warrant, which must be minutely described.

3d. The filing of the caveat in the General Land Office, and the advertisement of the loss being only preliminary steps towards the observance of the regulations, the owner of the lost warrant must file in the Pension Office as soon after the discovery of the loss as practicable, his declaration under oath, duly authenticated. setting forth fully and distinctly the time, place, and circumstances of the loss, and, if he be the original warrantee, that he never sold, assigned, nor voluntarily parted with his right to the warrant in question.

4th. In cases where a reissue of a warrant is sought on the ground of the non-reception of the original warrant, the agent or person to whom it was sent must unite with the warrantee or make a separate affidavit as to its non-reception.

5th. If the applicant for the reissue be not the person to whom the warrant was issued, but claims to be the owner thereof by purchase for a valuable consideration, he must give the name and residence of the warrantee, the name and residence of the person of whom he bought it, and, as far as he may know, or can ascertain, the names and residences of each of the several parties through whom the title of the warrant descended to him from the original warrantee, and adduce satisfactory evidence in proof of each and all his statements in reference thereto.

6th. The identity of the applicant must be satisfactorily established, and the credibility of each and every affiant must be duly certified by the magistrate administering the oaths, and his official character and signature must be verified by the proper officer under his seal of office.

7th. The Pension Office will also, for the space of three months, advertise the alleged loss of the original warrant and the pendency before it of the application for its reissue in the "Constitution" newspaper, published at the seat of government; and no warrant will be reissued under the foregoing act until after the expiration of three months from the date of the filing of the petition in this office, and not then if it shall appear that the original warrant is in existence.

STATE SELECTIONS.

The directions in the following circular of the General Land Office have reference, especially, to Selections for Railroad purposes. But, except the affidavits, are equally applicable to State Selections under other grants.

Each list should state specifically the grant under which the land is claimed, and bear on its face reference to the act conferring the grant by the date of its approval, and be properly verified by the signature of the selecting agent. All of the necessary forms for State or Corporation Selection lists, and for the proper verification of the same, may be obtained at the general or at the local land offices.

GRANTS FROM CONGRESS TO STATES AND CORPORATIONS.

DEPARTMENT OF THE INTERIOR,
GENERAL LAND OFFICE,
January 24th, 1867.

GENTLEMEN:

By the first section of the act of Congress approved July 1st, 1864, Statutes 1863–64, page 335, chap. 196, it is provided that from and after the passage of *that* act, "in the location of lands by States and corporations, under grants from Congress for railroads and other purposes (except for agricultural colleges), the Registers and Receivers of the Land Offices for the several States and Territories, in the districts where such lands may be located, for their services therein, shall be entitled to receive a fee of one dollar for each final location of one hundred and sixty acres, to be paid by the State or corporation making such location; the same to be accounted for in the same manner as fees and commissions on warrants and pre-emption locations, with limitations as to maximums of salary prescribed by existing laws, in accordance with such instructions as shall be given by the Commissioner of the General Land Office."

1st. Under this law the Registers and Receivers are *each* entitled to receive a fee of one dollar for each final location of one hundred and sixty acres, or any quantity approximate thereto, when the deficit is less than forty acres.

2d. When the several quantities shall have been definitely ascertained by you to inure to the grant, as hereinafter prescribed, the fees will then be due thereon.

3d. The State through its grantee, or the grantee, as the case may be, is required to file with the Register and Receiver of the proper land office descriptive lists of the tracts of land claimed as inuring under the grant, within sections of miles each, along the line of route on both sides thereof, to be dated and verified by the signature of the selecting agent.

For agent's certificate to be attached to each list, see Form A.

The party appearing as the agent of the grantee must file with the Register and Receiver written and satisfactory evidence, under seal, showing his authority to act in the premises.

In the preparation of the descriptive lists, the Register and Receiver will afford the agent all reasonable facilities, taking care, however, not to interrupt the current public business.

The lists must be carefully and critically examined by the Register and Receiver; their accuracy tested by the plats and records of their office. When so examined and tested, and found correct in all respects, to be a *final location;* and you will, on the payment of the requisite fees to the Receiver, so certify at the foot of each list, according to Form B.

After such lists have been examined, and you have attached your certificate thereto, the same will be consecutively numbered, commencing with No. 1, for each railroad or separate grant. Upon the payment of the fees and certification of the lists by you, the Register will post the selections in the Tract Book, after the following manner:

"Selected , 186 , by A. B., agent for the Railroad Co., act , list No. ;" and on the plats he will mark the tracts so selected " R. R."

After the selections are properly posted and marked on the plats, the lists will be transmitted to this office, accompanied by the evidence of the agent's appointment.

4th. The fees will be due in all cases where the service may have been rendered *subsequent* to the passage of said act of 1864.

5th. The Receiver will account for the fees thus paid in his monthly and quarterly accounts, specially setting forth in the same the particular case or cases on which such fees had accrued, giving the name of road, number and date of the list of selections for which they had been paid.

6th. By joint resolution No. 10 of January 30th, 1865, "mineral lands" are not embraced in the grants made at the 1st session of the 38th Congress, unless otherwise specially provided in the act or acts making the grants.

PACIFIC RAILROAD.

Acts approved July 1st, 1862, and July 2d, 1864.

7th. By section 21 of the latter act, these companies are required to pay cost of "surveying, selecting, and conveying" the lands, *in addition* to the Register and Receiver's fees exacted by the act of July 1st, 1864, before mentioned. This cost of surveying and conveying is, by the decision of the Secretary of the Interior of November 8th, 1866, limited to the lands granted by act July 2d, 1864. Therefore, the "cost" will be assessed and collected on the lands outside of ten miles and within twenty miles from the line of the road, where the grant is under both acts.

To ascertain the cost of "surveying," which includes both *surveying in the field* and *office work*, the company will apply to the Surveyor-General of the State or Territory in which the lands are situated. Upon ascertaining the sums due for surveying and office work for the "section or sections of road" for which selections have been or are to be made, a deposit of those sums must be made, to the credit of the Treasurer of the United States, with an authorized depositary. The duplicate of deposit must be filed with the Surveyor-General; whereupon he will transmit to the Register and Receiver of the proper land office his certificate of such payment having been made, specifying how much was for surveying and how much for office work, as per Form C.

The Surveyor-General's certificate, together with the triplicate certificate of deposit and the evidence of the agent's appointment, must accompany the lists of selection when transmitted by you to this office.

8th. Herewith is a form of title-page to be prefixed to the Lists of Selection.

Let me here call your special attention to the necessity of *great care* in the examination and testing of these lists, so that all conflicts may be avoided and improper selections be excluded, and that the verified schedules may be absolutely accurate, thus avoiding embarrassment and delay to all concerned.

9th. Pacific Railroad act, July 2d, 1864.—It is provided in section 4 that the word "mineral," when it occurs in that act, shall not be held to include iron and coal. Therefore, iron and coal lands are subject to selection by the Pacific Railroads; but all other minerals are expressly excluded from the grant, and must necessarily be so from all selections you may certify to this office. When the verified lists are received at the General Land Office, prepared and certified as above required, such definite action as the law requires will be here taken, with the view to invest the grantee with a complete title.

These instructions will supersede those of May 30th, 1866, Circular No. 9. The forms attached hereto, which are made a part hereof, will be followed in certifying to maps and lists, where the same may be applicable.

You will please acknowledge the receipt of this circular, giving the date of reception.

Jos. S. Wilson, *Commissioner.*

To the Registers and Receivers
of the U. S. Land Offices.

Department of the Interior,
Washington, D. C., January 29th, 1867.

The foregoing rules are approved.

O. H. Browning, *Secretary.*

PRE-EMPTIONS.

The act of September 4th, 1841, grants to settlers upon the public lands a preference right to purchase, known as the pre-emption right.

The individual claiming the benefits of said act must be—

1st. A citizen of the United States, or have filed his declaration of intention to become a citizen.

2d. Either the head of a family, or a widow, or a single man over the age of twenty-one years.

3d. An inhabitant of the tract sought to be entered, upon which, in person, he has made a settlement and erected a dwelling-house.

4th. The tract claimed to be in compact form, and consist of a regular quarter section or other legal subdivisions of sections.

5th. Claims initiated prior to the surveys will be adjusted by the lines of the surveys when made, in such manner as to cover the improvements of the claimants.

If the lands are surveyed at the date of settlement, only one person on a quarter section is protected by this law, and that is the one who made the *first settlement,* provided he shall have conformed to the other provisions of the law.

A person who has once *availed himself* of the provisions of this act cannot, at any future period, or at any other land office, acquire another right under it.

No person who is the proprietor of *three hundred and twenty acres* of land in any State or Territory of the United States, is entitled to the benefits of this act.

No person who shall *quit* or *abandon his residence* on his *own land* to reside on the public land in the *same State* or *Territory*, is entitled to the benefits of this act.

No pre-emption right exists by reason of a settlement on and inhabitancy of a tract, unless at the date of such settlement the Indian title thereto had been extinguished.

The approval of the plat is the evidence of the legality of the survey; but in accordance with the spirit and intent of the law, and for the purpose of bringing the settler within its provisions, the land is to be construed as surveyed when the requisite lines are run on the field and the corners established by the deputy surveyor.

No assignments or transfer of pre-emption rights are recognized. The patents must issue in the name of the claimants shown to be entitled to the land under the law.

The following description of lands are exempted from the operation of this act:

First. Lands included in any reservation or upon which are situated any known salines.

Second. Mineral lands.

Third. Lands selected for or upon which is situated any town or city.

Persons claiming the benefit of the pre-emption act are required to file in the local land office duplicate affidavits, as required by the law, and to furnish proof by one or more disinterested witnesses of the facts necessary to establish a full compliance with the requirements of the statute, and clearly to satisfy the department of the actual *bona fide* purpose of the pre-emptor to reside upon and cultivate the land claimed.

Forms for pre-emption affidavits and of proof will be found at all the local United States Land Offices, where the application and proof must be made.

Careful attention to the foregoing brief directions, and compliance with the provisions of the law, will enable parties to avoid conflicts and procure titles without delay.

FORM A.

No. LAND OFFICE AT

REPORT *of proposed selections of lands for School purposes under the provisions of the act of Congress of 20th May, 1826, entitled "An act to appropriate lands for the support of schools in certain townships and fractional townships not before provided for."*

TOWNSHIPS FOR WHICH THE SCHOOL SELECTION IS TO BE MADE.		CONTENTS OF TOWNSHIPS ENTITLED TO				LANDS SELECTED FOR SCHOOL PURPOSES				
		One Section.	¾ of a Section.	½ of a Section.	¼ of a Section.					
Number of Townships.	Number of Range.	Quantity more than 17,280 acres.	Quantity more than 11,520 acres, but less than 17,280 acres.	Quantity more than 5760 acres, but less than 11,520 acres.	Quantity more than 640 acres, but less than 5760 acres.	Section or part.	Number of Section.	Number of Township.	Number of Range.	Quantity.

FORM C.

LIST No. *Exhibiting the tracts of public land situated in the district of lands subject to sale at which have been selected for the State of under the 8th section of the act of Congress, approved 4th September, 1841, entitled "An act to appropriate the proceeds of the sales of the public lands and to grant pre-emption rights."*

Date of the filing of the list in the Register's Office, being the date the selection takes effect.	DESCRIPTON OF THE LAND.				AREA OF THE TRACTS.		This column exhibits the actual area of the tracts which contain 320 acres and upwards, and also cases in which the tracts, though containing a *less* quantity, are taken as equivalent to 320 acres. NOTE.—The total of this column is the *whole area* of the selections by this list.	REMARKS.
	Parts of Sections.	Section.	Township.	Range.	Acres.	100ths		

EDUCATIONAL GRANTS.

1st. For the support of primary schools, the grant is made of the land in place, being in the older land States, the section 16 in each township; but latterly the policy of the government has been liberalized, and in each township sections 16 and 36 are reserved for the endowment of primary schools. The lands cannot be disposed of by the territorial government, but are held in reserve until the organization of the State and its admission into the Union; when the title to the granted sections in place is vested in the State, and may be disposed of by the State government, the proceeds of sale forming the principal of a primary school fund, the interest of which may be annually applied to the support of public schools.

But in fractional townships, in which there is no granted section, its place being covered by the waters of the ocean, lakes, or rivers, the State under the act of May 20th, 1826, is entitled to indemnity, in the proportions shown in the following form,* in which the indemnity selection list is to be made out. The same to be signed by the State selecting agent, and verified by the local United States land officers of the district in which the lands are situated.

The act of February 26th, 1859, authorizes persons who settle upon the 16th and 36th sections prior to the survey to preempt the same, and in such cases allows indemnity to the State, the following being the form for selecting indemnity, under this provision of the statute:

Grants for seminary or university purposes have been made to each State to the amount of seventy-two sections. These lands may be selected by the Territorial government, in which case they are held in reserve until the admission of the State, when the title is consummated by certified list prepared by the General Land Office, and approved by the Secretary of the Interior.

Grants for agricultural colleges in States containing a sufficient quantity of public lands to satisfy the grant, may be

---

* See opposite page

FORM B.

LIST No. **OF INDEMNITY SCHOOL SELECTIONS**

Exhibiting *the tracts of Public Lands, situated in the district of lands subject to sale at* ***, which have been selected for the support of schools in certain townships and fractional townships where the full amount of school lands to which such townships are entitled have not been granted in place, and which selections are provided for in the act of Congress approved February 26th, 1859, entitled "An act to authorize settlers upon Sixteenth and Thirty sixth sections, who settled before the surveys of the Public Lands, to pre-empt their settlements."***

Townships Containing—								Portions of sections sixteen and thirty-six remaining in place in part satisfaction of school lands to which townships are entitled.			Deficiency in school lands of townships.	Selections to supply the deficiency in school lands to which townships are entitled.					Remarks as to cause of deficiency in school lands.
More than 17,280 acres. Entitled to 1280 acres.		*More than 11,520 acres, and less than 17,280 acres.* Entitled to 960 acres.		*More than 5760 acres, and less than 11,520 acres.* Entitled to 640 acres.		*More than 640 acres, and less than 5760 acres.* Entitled to 320 acres.											
Town.	Range.	Town.	Range.	Town.	Range.	Town.	Range.	*Description.*	*Sec.*	**Area.*	**Area.*	*Description.*	*Sec.*	*Town.*	*Range.*	*Area.*	

State of , *County of* .

We do hereby recommend the above selections, and agree to accept the same in lieu of the deficiency in school lands in certain townships and fractional townships, as shown in the foregoing list.

....

....

Land Office at 18 .

We do hereby certify that the foregoing list was filed at this Office on the day of , 18 , and that the selections are correct; and that there is no adverse interest conflicting therewith.

...., *Register.*

...., *Receiver.*

* Note No. 1.—The aggregate of columns numbers 11 and 12 must equal the exact area of school land to which township is entitled.

* Note No. 2.—No Indemnity School Selections are admissible by law for townships covered wholly by a private claim, or covered wholly or in part by an Indian reservation. Neither can Indemnity be taken by selections of mineral lands or lands which have been reserved.

selected by the State, and upon the list being approved by the General Land Office and the Secretary of the Interior, the title vests in the State.

SALT SPRING LANDS.

Saline reservations are made in each Territory; and upon becoming a State and being admitted into the Union, these reservations to the amount of seventy-two sections have been granted to each State.

INTERNAL IMPROVEMENTS.

By the act of September 4th, 1841, 500,000 acres of the public lands within the State are granted to each State for general works of Internal Improvements. Form C, page 191, has been prescribed in selecting this class of Lands. The following application and certificate are to be attached to the selections when made.

State of 186 .

I hereby apply in behalf of the State of
for the tracts described in this list, as being selected for said State under the 8th section of the act of 4th September, 1841.

Agent.

Land Office at 18 .

I hereby certify that the foregoing list was filed in this office on the and that the selections are correct, and that no valid conflicting right is known to exist.

Register.

MINING CLAIMS

UNDER

THE ACT OF CONGRESS APPROVED JULY 26, 1866.—U. S. STATUTES, PAGE 251, CHAPTER CCLXII.

DEPARTMENT OF THE INTERIOR,
General Land Office, January 14th, 1867.

GENTLEMEN: Herewith will be found the act of Congress approved 26th July, 1866, "granting the right of way to ditch and canal owners over the public lands, and for other purposes."

By the first section of this act all the mineral lands of the United States, surveyed and unsurveyed, are laid open to "all citizens of the United States, and to those who have declared their intention to become such, subject to statutory regulations," and also "to the local customs or rules of miners in the several mining districts not in conflict with the laws of the United States."

It therefore becomes your duty, *in limine*, to acquaint yourselves with the local mining customs and usages in the district in which you may be called upon to do those official acts which are required by law, whether the same are reduced to authentic written form, or are to be ascertained by the testimony of intelligent miners, which you are to obtain as occasion may require and justify, in acting upon individual claims, a perfect record whereof is to be carefully taken and preserved by the Register and Receiver, and to be accompanied by a diagram or plat fixing the out boundaries of the district in which such customs and usages exist.

The second section of the act declares that "whenever any person or association of persons claim a vein or lode of quartz or other rock in place, bearing gold, silver, cinnabar, or copper, having previously occupied and improved the same according to the local custom or rules of miners in the district where the same is situated, and having expended in actual labor and improvements thereon an amount of not less than one thousand dollars, and in regard to whose possession there is no controversy or opposing claim, it shall and may be lawful for said claimant, or association of claimants, to file in the local land office a diagram of the same, so extended *laterally* or otherwise, as to conform to the local laws, customs, and rules of miners, and to enter such tract and receive a patent therefor, granting such mine, together with the right to follow such vein or lode with its dips, angles, and variations, to any depth, although it may enter the land adjoining, which land adjoining shall be sold subject to this condition."

Mining claims may be entered at any district land office in the United States under this law by any person, or association of persons, corporate or incorporate. In making the entry, however, such a description of the tract must be filed as will indicate the vein or lode, or part or portion thereof claimed, together with a diagram representing, by reference to some natural or artificial monument, the position and location of the claim and the boundaries thereon, so far as such boundaries can be ascertained.

First. In all cases the number of feet in length claimed on the vein or lode shall be stated in the application filed as aforesaid, and the lines limiting the length of the claim shall, also, in all cases be exhibited on the diagram, and the course or direction of such end lines, when not fixed by agreement with the adjoining claimants, nor by the local customs or rules of the miners of the district, shall be drawn at right angles to the ascertained or apparent general course of the vein or lode.

Second. Where, by the local laws, customs, or rules of miners of the district, no surface ground is permitted to be occupied for mining purposes except the surface of the vein or lode, and the walls of such vein or lode are unascertained, and the lateral extent of such vein or lode unknown, it shall be sufficient, after giving the description and diagram aforesaid, to state the fact that the extent of such vein or lode cannot be ascertained by actual measurement, but that the said vein or lode is bounded on each side by the wall of the same, and to estimate the amount of ground contained between the given end lines and the unascertained walls of the vein or lode; and in such case the patent will issue for all the land contained between such end lines and side walls, with the right to follow such vein or lode, with all its dips, angles, and variations, to any depth, although it may enter the land adjoining: *Provided*, The estimated quantity shall be equal to a horizontal plane bounded by the given end lines, and the walls on the sides of such vein or lode.

Third. Where, by the local laws, customs, or rules of miners of the district, no surface ground is permitted to be occupied for mining purposes, except the surface of the vein or lode, and the walls of such vein or lode are ascertained and well known, such wall shall be named in the description, and marked on the diagram, in connection with the end lines of such claims.

Fourth. Where, by the laws, customs, or rules of miners of the district, a given quantity of surface ground is fixed for the purpose of mining or milling the ore, the aforesaid diagram and description in the entry shall correspond with and include so much of the surface as shall be allowed by such laws, customs, or rules for the purpose aforesaid.

Fifth. In the absence of uniform rules in any mining district limiting the amount of surface to be used for mining purposes, actual and peaceable use and occupation for mining or milling purposes shall be regarded as evidence of a custom of miners authorizing the same, and the ground so occupied and used in connection with the vein or lode, and being adjacent thereto, may be included within the entry aforesaid, and the diagram shall embrace the same as appurtenant to the mine.

Where the claimant or claimants desire to include within their entry and diagram any surface ground beyond the surface of the vein, it shall be necessary, upon filing the application, to furnish the Register of the land office with proof of the usage, law, or custom under which he or they claim such surface ground, and such evidence may consist either of the written rules of the miners of the district or the testimony of two credible witnesses to the uniform custom or the actual use and occupation as aforesaid, which testimony shall be reduced to writing by the Register and Receiver, and filed in the Register's office, with the application, a record thereof to be made as contemplated under the first head in the foregoing.

By the *third section* of the act it is required that upon the filing of the diagram as provided in the second section, and posting the same in a conspicuous place on the claim, with notice of intention to apply for a patent, the Register shall publish a notice of the same in a newspaper nearest the location of said claim, which notice shall state name of the claimant, name of mine, names of adjoining claimants on each end of the claim, the district and county in which the mine is situated, informing the public that application has been made for a patent for same, the Register also to post such notice in his office for ninety days.

Thereafter, should no adverse claim have been filed, and satisfactory proof should be produced that the Diagram and Notice have been posted in the manner and for the period stipulated in the statute, it will become the duty of the Surveyor-General, upon application of the party, to survey the premises, and make plat thereof, indorsed with his approval, designating the number and description of the location, the value of the labor and improvements, and the character of the vein exposed. As preliminary to the survey, however, the Surveyor-General must estimate the expense of surveying, platting, and ascertain from the Register the cost of the publication of notice, the amount of all of which must be deposited by the applicant for survey with any assistant United States Treasurer or designated depositary, in favor of the United States Treasurer, to be passed to the credit of the fund created by "individual depositors for the surveys of the public lands." Duplicate certificates of such deposits must be filed with the Surveyor-General for transmission to this office, as in the case of deposits for surveys of public lands under the 10th section of the act of Congress approved May 30th, 1862, and joint resolution of July 1st, 1864.

After the survey thus paid for shall have been duly executed, and the plat thereof approved by the Surveyor-General, designating the number and the description of the location, accompanied by his official certificate of the value of the labor and improvements and character of the vein exposed, with the testimony of two or more reliable persons cognizant of the facts on which his certificate may be founded as to the value of the labor and improvements, the party claiming shall file the same with the Register and Receiver, and thereupon pay to the said Receiver five dollars per acre for the premises embraced in the survey, and shall file with those officers a triplicate certificate of deposit showing the payment of the cost of survey, plat, and notice, with satisfactory evidence, which shall be the testimony of at least two credible witnesses, that the diagram and notice were posted on the claim for a period of ninety days, as required by law and as contemplated in the foregoing. Thereupon it shall be the duty of the Register to transmit to the General Land Office said plat, survey, and description, with the proof indorsed as satisfactory by the Register and Receiver, so that a patent may issue if the proceedings are found regular, but neither the plat, survey, description, nor patent shall issue for more than one vein or lode.

The unity of the surveying system is to be maintained by extending over the mining districts the rectangular method, at least so far as township lines are concerned.

The contemplated surveys of the mineral lands will be made by district deputies, under contracts, according to the mode adopted in the survey of the public lands and private land claims, embracing in them all such veins or lodes as may be called for by claimants entitled to have them surveyed.

In consideration of the very limited scope of surveying involved in each mining claim, the per mileage allowed by law may not be adequate to secure the services of scientific surveyors, and hence the necessity of resorting to a per diem principle, it being the most equitable under the circumstances.

The Surveyor-General is therefore hereby authorized to commission resident mineral surveyors for different districts where, isolated from each other, and absolutely inconvenient for one surveyor promptly to attend to the several calls for surveying in such localities, the compensation not to exceed ten dollars per diem, including all expenses incident thereto. Such surveyors shall enter into bonds of $10,000 for the faithful performance of their duties in the survey of such claims as the Surveyor-General may be required to execute in pursuance of the aforesaid law and these instructions.

The *fourth section* contemplates the location and entry of a mine upon unsurveyed lands, stipulating for the surveys of public lands to be adjusted to the lines of the claims, according to the location and possession and plat thereof. In surveying such claims, the Surveyor-General is authorized to vary from the rectangular form to suit the circumstances of the country, local rules, laws, and customs of miners. The extent of the locations made from and after the passage of the act shall, however, not exceed two hundred feet in length along the vein for each locator, with an additional claim for discovery to the discoverer of the lode, with the right to follow such vein to any depth, with all its dips, variations, and angles, together with a reasonable quantity of surface for the convenient working of the same as fixed by local rules: *Provided*, No person may make more than one location on the same lode, and no more than three thousand feet shall be taken in any one claim by any association of persons.

The deputy surveyors should be scientific men, capable of examining and reporting fully on every lode they will survey, and to bring in duplicate specimens of the ore, one of which you will send to this office and the other the Surveyor-General will keep to be ultimately turned over with the surveying archives to the State authorities.

The surveyors of mineral claims, whether on *surveyed* or *unsurveyed* lands, must designate those claims by a progressive series of numbers, beginning with No. 37, so as to avoid interference in that respect with the regular *sectional* series of numbers in each township; and shall designate the four corners of each claim, where the side lines of the same are known, so that such corners can be given by either trees, if any are found standing in place, or any corner rocks exist in place, or posts may be set diagonally and deeply imbedded, with four sides *facing* adjoining claims, sufficiently flattened to admit of inscriptions thereon; but where the corners are unknown, it will be sufficient to place a well built solid mound at each end of the claim. The beginning corner of the claim nearest to any corners of the public surveys is to be connected by course and distance, so as to ascertain the relative position of each claim in reference to township and range when the same have been surveyed; but in those parts of the surveying district where no such lines have as yet been extended, it will be the duty of Surveyors-General to have the same surveyed and marked, at least so far as standard and township lines are concerned, at the per mileage allowed, so as to embrace the mineral region, and to connect the nearest corners of the mineral claims with the corners of the public surveys.

Should it, however, be found impracticable to establish independent base and meridian lines, or to extend township lines over the region containing mineral claims required to be surveyed under the law, then, and in that case, you will cause to be surveyed in the first instance such a claim, the initial point of which will start either from a confluence of waters or such natural and permanent objects as will unmistakably identify the point of the beginning of the survey of the claim upon which other surveys will depend.

Sec. 5. Provides that in cases where the laws of Congress are silent upon the subject of rules for working mines, respecting easements, drainage, and other necessary means to the complete development of the same, the local Legislature of any State or Territory may provide them, and in order to embody such enactments into patents, you are directed to communicate any such laws to this office.

Sec. 6. Should adverse claimants to any mine appear before the approval of the survey, all further proceedings shall be stayed until a final settlement and adjudication are had in the courts of the rights of possession to such claim, except where the parties agree to settlement or a portion of the premises is not in dispute, when a patent may issue as in other cases.

Sec. 7. Provides for such additional land districts as may be necessary.

Sec. 8. For the right of way.

Sec. 9. For the protection of rights to the use of water for mining, agricultural, manufacturing, or other purposes, for the right of way for the construction of ditches and canals; and makes parties constructing such work (after the passage of this act) to the injury of settlers, liable in damages.

Sec. 10. Homesteads made prior to the passage of this act by citizens of the United States, or persons who have declared their intention to become citizens, but on which lands no valuable mines of gold, silver, cinnabar, or copper have been discovered, are protected, so that settlers or owners of such homesteads shall have a right of pre-emption thereto, in quantity not to exceed one hundred and sixty acres, at $1.25 per acre, or to avail themselves of the Homestead act and acts amendatory thereof.

Sec. 11. Stipulates that upon the survey of the lands in question, the Secretary of the interior may set apart such portions as are clearly agricultural, and thereafter subjects such agricultural tracts to pre-emption and sale as other public lands.

In order to enable the department properly to give effect to this section of the law, you will cause your deputy surveyors to describe in their field notes of surveys, in addition to the data required to be noted in the printed Manual of Surveying Instructions on pages 17 and 18, the agricultural lands, and represent the same on township plats by the designation of "Agricultural lands."

It is to be understood that there is nothing obligatory on claimants to proceed under this statute, and that where they fail to do so, there being no adverse interest, they hold the same relations to the premises they may be working which they did before the passage of this act, with the additional guarantee that they possess the right of occupancy under the statute.

The foregoing presents such views as have occurred to this office in considering the prominent points of the statute, and will be followed by further instructions as the rulings in actual cases, and experience in the administration of the statute may from time to time suggest.

Very respectfully, your obedient servant,

Jos. S. Wilson, *Commissioner.*

To the United States Registers and Receivers
and Surveyors-General.

AN ACT granting the right of way to ditch and canal owners over the public lands and for other purposes.

Be it enacted by the Senate and House of Representatives of the United States of America in Congress assembled, That the mineral lands of the public domain, both surveyed and unsurveyed, are hereby declared to be free and open to exploration and occupation by all citizens of the United States, and those who have declared their intention to become citizens, subject to such regulations as may be prescribed by law, and subject also to the local customs or rules of miners in the several mining districts, so far as the same may not be in conflict with the laws of the United States.

SEC. 2. *And be it further enacted,* That whenever any person or association of persons claim a vein or lode of quartz, or other rock in place, bearing gold, silver, cinnabar, or copper, having previously occupied and improved the same according to the local custom or rules of miners in the district where the same is situated, and having expended in actual labor and improvements thereon an amount of not less than one thousand dollars, and in regard to whose possession there is no controversy or opposing claim, it shall and may be lawful for said claimant or association of claimants to file in the local land office a diagram of the same, so extended laterally or otherwise as to conform to the local laws, customs, and rules of miners, and to enter such tract and receive a patent therefor, granting such mine, together with the right to follow such vein or lode with its dips, angles, and variations to any depth, although it may enter the land adjoining, which land adjoining shall be sold subject to this condition.

SEC. 3. *And be it further enacted,* That upon the filing of the diagram as provided in the second section of this act, and posting the same in a conspicuous place on the claim, together with a notice of intention to apply for a patent, the Register of the land office shall publish a notice of the same in a newspaper published nearest to the location of said claim, and shall also post such notice in his office for the period of ninety days; and after the expiration of said period, if no adverse claim shall have been filed, it shall be the duty of the Surveyor-General, upon application of the party, to survey the premises and make a plat thereof, indorsed with his approval, designating the number and description of the location, the value of the labor and improvements, and the character of the vein exposed; and upon the payment to the proper officer of five dollars per acre, together with the cost of such survey, plat, and notice, and giving satisfactory evidence that said diagram and notice have been posted on the claim during said period of ninety days, the Register of the land office shall transmit to the General Land Office said plat, survey, and description, and a patent shall issue for the same thereupon. But said plat, survey, or description shall in no case cover more than one vein or lode, and no patent shall issue for more than one vein or lode, which shall be expressed in the patent issued.

SEC. 4. *And be it further enacted,* That when such location and entry of a mine shall be upon unsurveyed lands, it shall and may be lawful, after the extension thereto of the public surveys, to adjust the surveys to the limits of the premises according to the location and possession and plat aforesaid; and the Surveyor-General may, in extending the surveys, vary the same from a rectangular form to suit the circumstances of the country and the local rules, laws, and customs of miners: *Provided*, That no location hereafter made shall exceed two hundred feet in length along the vein for each locator, with an additional claim for discovery to the discoverer of the lode, with the right to follow such vein to any depth, with all its dips, variations, and angles, together with a reasonable quantity of surface for the convenient working of the same, as fixed by local rules: *And provided further*, That no person may make more than one location on the same lode, and not more than three thousand feet shall be taken in any one claim by any association of persons.

SEC. 5. *And be it further enacted*, That as a further condition of sale, in the absence of necessary legislation by Congress, the local Legislature of any State or Territory may provide rules for working mines involving easements, drainage, and other necessary means to their complete development; and those conditions shall be fully expressed in the patent.

SEC. 6. *And be it further enacted*, That whenever any adverse claimants to any mine located and claimed as aforesaid shall appear before the approval of the survey, as provided in the third section of this act, all proceedings shall be stayed until a final settlement and adjudication in the courts of competent jurisdiction of the rights of possession to such claim, when a patent may issue as in other cases.

SEC. 7. *And be it further enacted*, That the President of the United States be, and is hereby authorized to establish additional land districts, and to appoint the necessary officers under existing laws, wherever he may deem the same necessary for the public convenience in executing the provisions of this act.

SEC. 8. *And be it further enacted*, That the right of way for the construction of highways over public lands, not reserved for public uses, is hereby granted.

SEC. 9. *And be it further enacted*, That whenever by priority of possession, rights to the use of water for mining, agricultural, manufacturing, or other purposes, have vested and accrued, and the same are recognized and acknowledged by the local customs, laws, and the decisions of courts, the possessors and owners of such vested rights shall be maintained and protected in the same; and the right of way for the construction of ditches and canals for the purposes aforesaid is hereby acknowledged and confirmed: *Provided, however*, That whenever, after the passage of this act, any person or persons shall, in the construction of any ditch or canal, injure or damage the possession of any settler on the public domain, the party committing such injury or damage shall be liable to the party injured for such injury or damage.

SEC. 10. *And be it further enacted*, That wherever, prior to the passage of this act, upon the lands heretofore designated as mineral lands, which have been excluded from survey and sale, there have been homesteads made by citizens of the United States, or persons who have declared their intention to become citizens, which homesteads have been made, improved, and used for agricultural purposes, and upon which there have been no valuable mines of gold, silver, cinnabar, or copper discovered, and which are properly agricultural lands, the said settlers or owners of such homesteads shall have a right of pre-emption thereto, and shall be entitled to purchase the same at the price of one dollar and twenty-five cents per acre, and in quantity not to exceed one hundred and sixty acres; or said parties may avail themselves of the provisions of the act of Congress approved May twenty, eighteen hundred and sixty-two, entitled "An act to secure homesteads to actual settlers on the public domain," and acts amendatory thereof.

SEC. 11. *And be it further enacted,* That upon the survey of the lands aforesaid, the Secretary of the Interior may designate and set apart such portions of the said lands as are clearly agricultural lands, which lands shall thereafter be subject to pre-emption and sale as other public lands of the United States, and subject to all the laws and regulations applicable to the same.

Approved July 26th, 1866.

THE COAL AND MINERAL RESOURCES

OF THE

UNITED STATES.

The following statements in regard to the mineral wealth of the country, are extracted from advance sheets of the Commissioner's Annual Report of the General Land Office, for 1867, for which the writer makes due acknowledgment.

From 1830 until 1861, mining was regularly carried on in Virginia, and from $50,000 to $100,000 annually received at the mint from that State, the whole amount deposited up to the year 1866 being $1,570,182 82, the first deposit of $2500 having been made in 1829. The gold belt in Virginia is from fifteen to twenty miles in width, and thus far developed chiefly in the counties of Fauquier, Culpeper, Orange, Spottsylvania, Louisa, Fluvanna, Goochland, Buckingham, Campbell, and Pittsylvania.

Gold was known to exist in North Carolina before the commencement of the present century, a good-sized nugget having been found in Cabarrus county in 1799, and another afterwards, weighing twenty-eight pounds avoirdupois. In the same locality it is estimated that over a hundred pounds were collected prior to 1830, in pieces each over one pound in weight. In the adjoining counties lumps were found weighing from one to sixteen pounds. From 1804 to 1827 North Carolina furnished all the gold of the United States, amounting, according to the mint returns, to $110,000. Up to the year 1866 the state deposited at the mint $9,278,627 67. The counties in which mining has been conducted are Rockingham, Guilford, Davidson, Rowan, Cabarrus, Rutherford, and Mecklenburg. Previous to 1825 the metal had been obtained from washings, but in that year auriferous vein stones were discovered and six hundred and twenty five ounces of gold obtained by rock mining, after which other leads were found in most of the counties above named.

In 1829 $3500 were deposited at the mint from South Carolina, and from 1830 to 1861 mining was prosecuted in that state with varying success In 1852 the Dorn mine was opened in the Abbeville district, and in a little more than a year produced $300,000 worth of gold by the aid of a single Chilian mill worked by two mules. The total deposit from this state amounts to $1,353,663 98. The whole northwestern part of South Carolina contains gold, but the districts in which it has been mainly developed are Abbeville, Pickens, Spartanburg, Union, York, and Lancaster.

In 1830 $212,000 were received from Georgia as the first contribution of its mines, which from that date to 1861 yielded a product of $6,971,681 50. The whole of the state lying along the base of the Blue Ridge has been found more or less auriferous, but the counties in which mining has been principally conducted are Carroll, Cobb, Cherokee, Lumpkin, and Habersham.

Gold has been found in Tennessee and Alabama, but the quantity has been small, the whole amount deposited from the former state since 1828 being only $81,406 75, and from the latter since 1838, $201,734 83.

Efforts are now being made to develop the quartz veins of the southern states with the aid of the improvements in mining found to be effective in California and elsewhere.

But the most important gold fields of the United States and of the world are found in the states and territories extending from the northern to the southern boundaries of the republic, and from the Pacific Ocean to the eastern spurs and outlines of the Rocky Mountains, embracing an area of more than a million of square miles.

OHIO.—The coal fields in the eastern and southeastern portions of this state cover an area of 12,000 square miles, extending through twenty counties, and embrace nearly one-third of the area of the whole state; it being estimated that the county of Tuscarawas alone is underlaid with an amount equal to eighty thousand millions of bushels. Iron ore of very superior quality for the finer castings is found in several counties in the southern bend of the Ohio, covering an area of 1200 square miles, and has already laid the foundation of a very extensive iron interest in the southern part of the state. In the northern part the furnaces are supplied with ore from the Lake Superior mines.

Large quantities of salt are manufactured for market.

Many oil wells have been sunk in the southeastern portion, and large quantities of oil have been exported.

In 1860, according to the estimates of the commissioner of statistics for the state, 50,000,000 bushels of coal were mined, and 2,000,000 bushels of salt manufactured. Ohio ranked next to Pennsylvania in the production of coal and pig iron, the latter state standing first in these industries. For the manufacture of salt Ohio stood third. The state has doubled its products and manufactures every ten years since 1840.

INDIANA.—The great coal field of Illinois extends into Indiana, covering in the western part an estimated area equal to 7700 square miles, or more than one-fifth part of the whole surface. On White River the seams are upwards of six feet thick. In other localities seams of eight feet in thickness are found. Some of the coal measures, it is estimated, are capable of yielding 50,000,000 bushels to the square mile. At Cannelton, on the Ohio, a bed of cannel coal is found from three to five feet in thickness, at an elevation of seventy feet above the river.

Besides coal, iron, limestone, marble, freestone, gypsum, and grindstones, slate of several varieties, clays useful in the arts, and some copper are found in the state.

ILLINOIS.—The Illinois coal field stretches from the Mississippi, near Rock Island, eastward toward Fox River, thence southeast through Indiana, and southward into Kentucky, occupying the greater part of Illinois, the southwestern portion of Indiana, and the northwestern part of Kentucky, measuring three hundred and seventy-five miles in length from northwest to southeast, and two hundred in width from St. Louis eastward—estimated to contain 1,277,500,000,000 tons of coal, sufficient to furnish an annual supply of 13,000,000 tons for nearly a hundred thousand years, being more than six times as large as all the coal fields of Great Britain, and embracing one-third of all the coal measures of North America.

The present annual product of the state is 1,500,000 tons, the amount increasing every year, and, as the coal is of good quality and easily mined, it is destined to become one of the most prominent interests of the state.

The great lead district of the Mississippi river occupies a portion of northwestern Illinois, southwestern Wisconsin, and northeastern Iowa, covering an area of about 1,000,000 acres, one-sixth of which lies in Illinois, in Jo Daviess county, which has furnished the entire lead product of the country for twenty years. A few mines in Wisconsin and Illinois have supplied and smelted 15,000 000 pounds a year.

Iron ore has been mined in Hardin county, on the Ohio, several furnaces being in operation. Valuable beds of the ore are reported between the Kaskaskia and the Mississippi; also in Union county and in the northern part of the state. Copper has been found in several counties; also marble, crystallized gypsum, quartz crystal, and silex for glass manufacture; salt also existing in the southern counties, while small quantities of gold and silver have been obtained in the lead district in the northwest corner of the state. Petroleum is found in the northeast part, zinc ore in the lead district in Jo Daviess, sulphur and chalybeate springs in Jefferson and other localities.

MICHIGAN.—The upper peninsula, rich in minerals, prominent among which is copper, is mostly of primitive geological character; the lower exclusively secondary. The copper deposits among the primary rocks of the northern peninsula are the richest in the world, the copper belt being one hundred and twenty miles long and from two to six miles wide. A block of several tons of almost pure copper, taken from the mouth of Ontonagon river, has been built into the wall of the Washington monument at the national capital. A mass weighing one hundred and fifty tons was uncovered in 1854 in the North American mine.

Isle Royale abounds in this mineral; one house in that district, during five and a half months of 1854, shipped over two millions of pounds, and in the nine years previous there were produced four thousand eight hundred and twenty-four tons. The yield of copper in the state has risen to an annual average of eight thousand tons, with promise of steady increase. The opening of the St. Mary's canal and the clearing of the entrance into Portage Lake have given fresh impetus to this branch of mining industry, which is becoming one of the most cherished interests of the state. Silver has been found in connection with the copper in the proportion of from twenty-five to fifty per cent. of the precious metal. Iron of superior quality has been discovered in a bed of slate from six to twenty-five miles wide, and one hundred and fifty long, extending into Wisconsin. In the production of this mineral in 1863, Michigan was second only to Pennsylvania, having produced two hundred and seventy-three thousand tons of ore. Bituminous coal is mined on an enlarging scale to meet the demand of manufactures. Salt also exists in quantities repaying the investment of capital.

The high prices lately prevailing have caused a rapid development of the salt fields around Saginaw, a basin some forty or fifty miles square, in which by boring some eight hundred feet an inexhaustible supply of brine is obtained, yielding eighty or ninety per cent. of salt.

WISCONSIN.—The mineral resources of the state are varied and valuable. The lead region of Illinois and Iowa extends over an area of 2140 square miles in Wisconsin, which compares with the other portions in the abundance and richness of the ores. In 1863 there were 848,625 pounds of lead received at Milwaukee. The completion of the southern Wisconsin railroad will raise the aggregate to 2,500,000 pounds. It is mingled with copper and zinc ores.

The iron region of Lake Superior presents within the limits of this state abundant deposits of great richness. Magnetic iron, plumbago, and the non-metallic earths abound. Copper deposits have also been developed, but as yet have only been worked to a limited extent. Beautiful marbles, susceptible of elaborate working, exist.

IOWA.—The mineral resources of Iowa are abundant and rich, the lead region of Illinois and Wisconsin extending into this state, the ore being found in large quantities, but lying deeper than on the east side of the Mississippi. Dubuque is the center of the Iowa lead region. From this point and Buena Vista, in 1853, were shipped 3,256,970 pounds of this mineral. Zinc and copper are found in the same localities in close association with it. Coal is abundant and accessible.

MISSOURI.—Missouri is richly endowed with mineral wealth. The iron region around Iron mountain and Pilot Knob is unsurpassed in the world for the abundance and purity of deposits.

On the Maramec river and in some other localities are found small quantities of lead.

Copper is found extensively deposited, being most abundant near the La Motte mines. It is also found with nickel, manganese, iron, cobalt, and lead, in combinations yielding from thirty to forty per cent. All of these metals, except nickel, exist in considerable quantities; also silver, in combination with lead ore and tin. Limestone, marble and other eligible building materials are abundant, especially north of the Missouri. The geological formations of the state are principally those between the upper coal measures and the lower silurian rocks. The drift is spread over a large surface; in the north, vast beds of bituminous coal, including cannel coal, exist on both sides of the Missouri river.

ARKANSAS.—The minerals of Arkansas are chiefly iron, coal, lead, zinc, manganese, gypsum, and salt. The coal embraces deposits of the anthracite, cannel, and bituminous varieties.

Gold is said to have been found in White county. Near Hot Springs is a quarry of novaculite, or oil-stone, superior to any other on the globe, inexhaustible in quantity and of great variety of fineness. There is manganese enough in the state to supply the world's demand. In zinc Arkansas ranks next to New Jersey. It has more gypsum than all the other states, and is equally well supplied with marble and salt. The lead ore is largely associated with silver. Lead mines were worked extensively during the civil war to answer military necessities.

MINNESOTA.—The mineral resources of the state are yet undeveloped. Copper has been found, but in most cases as a detritus carried away from its deposits imbedded in moving masses of boulders and drift. Yet around Lake Superior it is likely a considerable amount of this mineral will be produced.

Coal has not yet been discovered in quantities comparable to its development in the neighboring states. Lead gives promise of greater abundance. In the northeastern part large formations of gold and silver-bearing quartz, accompanied by still further developments of iron ore, were reported by the state geologist upon actual survey in 1855.

KANSAS.—The mineral deposits of Kansas are as yet imperfectly known; but sufficient has been developed incidentally to warrant the belief that the state has a liberal endowment of the useful minerals. The coal measures of the eastern portion are supposed to cover an extensive region. The upper stratum crops out in the eastern and middle counties. In Leavenworth, Osage, and Bourbon counties, in which alone the veins have been worked to any extent, the supply seems abundant and the quality superior. The Surveyor-General is of opinion from late surveys that the coal veins to the westward will be found of increasing abundance. West of this out-cropping of the coal strata is an irregular belt, from fifty to seventy-five miles wide, of permian rocks, in which are found salt springs, inexhaustible beds of gypsum, and inferior iron ore; platinum has also been discovered. Sand rock and lime rock crop out all over the state. Fine magnesian limestone of beautiful color has been quarried on the Kansas river near Fort Riley, and is now used in erecting public buildings.

NEW MEXICO.—Surveyor-General Clark, of New Mexico, in his report to the General Land Office, gives the following statement in regard to the mining interests of that territory:

Gold.—Since my last annual report gold has been discovered in the mountains, about twenty miles northeasterly from the town of Taos, in this territory. At last advices it was estimated there were four hundred men engaged in gold-washing in a district five by fifteen miles in extent. I have no reliable information as to the actual production.

The New Mexico Mining Company is increasing and perfecting its machinery and increasing the working force at the old placer, twenty-eight miles southeasterly from Santa Fé, and promises largely to increase the production of gold from those mines during the next year. Dr. Michael Steck, the present superintendent, gives the result of the reduction of sixty-three tons of quartz from the mines of this company at seventeen hundred and seven dollars and sixty-four cents, or an average of twenty-seven dollars and ten cents per ton, and says that the ore is abundant, cheaply mined, and convenient to the works. Many lodes rich in gold have been discovered in that vicinity, but no other mill has been erected, and they remain undeveloped.

The developments at Pinos Altos during the year fully justify all I have heretofore reported concerning the mineral wealth of that region. It is estimated that there are now more than one thousand persons engaged in gold-washing and in working the rich veins of gold-bearing quartz in that immediate vicinity.

I am indebted for the following information concerning these mines to Brevet Major-General James H. Carleton, United States army, who lately visited them:

The Pinos Altos Mining Company has a quartz-crushing mill of fifteen stamps now in operation at the town of Pinos Altos. The quartz worked by it is taken from the Pacific lode, and yields from eighty to one hundred and fifty dollars of gold per ton. The cost of mining and delivering the ore at the mill is estimated at eight dollars and fifty cents per ton, and of reducing it and separating the gold at three dollars. The mill has the capacity of reducing twenty tons of ore in twenty-four hours. General Carleton was informed by good authority that within a radius of six miles from the town of Pinos Altos there had been discovered at the time of his visit six hundred lodes of gold and silver ore, many of them prospecting as rich as the Pacific lode above mentioned. There is a scarcity of water for washing; but in the rainy season, in many of the ravines or gulches in the vicinity, there will be water, so that miners can wash five to six dollars per diem to the hand.

Silver.—Numerous veins of silver ore are reported to have been discovered during the year in the Sandia, Manzano, San Andros, Mimbres, and Organ mountains, but none have been worked sufficiently to prove their value or extent. The greater part of the gold bearing quartz in New Mexico yields also more or less silver; and as a rule, I believe the percentage of silver increases as the veins descend. Silver, therefore, promises ultimately to be the leading mining interest in this section of the Rocky mountains. There are no works in operation for the reduction of the silver ores.

Copper.—Copper seems to be a universal accompaniment of the precious metals in this section; traces of it are found in most of the veins of gold and silver ore. Lodes and deposits of copper ore are reported to have been discovered in the Taos, Jemez, Sandia, and Mimbres mountains. When, by the construction of railroads, cheap transportation shall be furnished to the people, copper mining will become an important branch of the industry of this territory. The silver and gold in much of the ore will more than pay for its transportation and reduction.

Coal.—Veins of bituminous coal have been found in the Raton, Sandia, and Jemez mountains, near the Peurco river, west of Albuquerque, and in the vicinity of Forts Craig, Stanton, Selden, and Bayard. Anthracite coal of a superior quality is also found near the Galisteo creek, about twenty miles south from Santa Fé. I have no doubt but that this valuable mineral exists in abundance throughout the territory, and can be made available to furnish cheap fuel for the operation of railroads, and for manufacturing and domestic uses.

Lead and Iron.—Lead and iron are very common minerals throughout the territory. Much of the lead has sufficient percentage of silver to pay for its separation ; but as yet there is little domestic demand for lead, and the cost of transportation to a foreign market would consume it; there is, therefore, none mined or smelted. For the same reasons the mountains of iron ore remain untouched by the manufacturer.

Salt.—Almost the entire amount of salt used in New Mexico is obtained from salt lakes on the plain, fifty to sixty miles east of the Rio Grande. The salt, crystallized by the evaporation of the water by the sun, is deposited upon the bottom of the lake, forming a crust several inches thick, and is shoveled thence directly into the wagons and dried by the sun. There are some impurities mixed with it, which give it a dark appearance, but when leached, or washed, it becomes white as snow. The supply seems inexhaustible.

Other Minerals.—In addition to those above enumerated, zinc, antimony, kaolin, and other minerals are known to exist, which, when the railroads shall reach this region, and the current of immigration turns in this direction, with its capital and industry, to develop and work the mines, will contribute largely to the general wealth.

COLORADO. — The Surveyor-General of Colorado reports as follows :

Gold and Silver.—Of the gold and silver one can form no idea of the wealth of the deposits in this territory, and as soon as a method of separating the different metals in a less expensive manner is adopted, large results will follow.

The mines have not fully recovered from the effects of the late depression, owing in a great measure to reckless speculation But I am convinced that when fully developed, the mines of Colorado will be found second to none in riches.

Coal.—Coal exists in large quantities, and has been traced and opened along the base of the mountains, and the indications are that an extensive basin exists underlying a large extent of territory eastward from the mountains. The quality is good. It makes an excellent gas and steam coal, and some of it could be used for smelting iron.

Iron.—Iron is found in abundance along the base of the mountains, and at some distance from them, and with abundance of coal found near it, will prove in time invaluable. As yet no effort has been made to any extent to work it, owing to the high price of labor.

CALIFORNIA.—The great and distinguishing feature of California is its unexampled mineral wealth. The first discoveries of gold were made in 1848, when $10,000,000 were taken from the mines, increasing to $10,000,000 in 1849, and upwards of $65,000,000 in 1853.

No returns are made of the quantity taken from the mines, and the mint records are the only official data existing upon the product for any portion of the Pacific coast. Various estimates have been made by mining engineers, bankers, and other intelligent and practical business men in San Francisco and elsewhere in California as to the total product of that state since 1848. These estimates vary from eight hundred millions to one billion. From the commencement of 1849 to the close of 1866 upwards of seven hundred and eighty-five millions have been manifested at San Francisco for exportation, all of which, with the exception of sixty-five millions, appears to have been the product of California. How large a portion of gold found its way out of the state without being manifested for exportation is, of course, a matter of conjecture, different authorities estimating it from one hundred millions to three hundred millions. But either estimate is sufficient to furnish an idea of the immensity of the mineral wealth of the state.

Silver mines in the state are comparatively inconsiderable, yet quantities of that metal are annually obtained by separating it from gold, with which it is, in small portions, generally united when taken from the mines. The quicksilver mines of California are among the most valuable, and have since their discovery materially contributed to the prosperity of the mining interests not only of California and the adjoining states, but also in Mexico and South America. All the useful metals, such as iron, lead, copper, tin, and zinc, exist in this region. Coal has been discovered in different localities, and marble, gypsum, and valuable building stones are abundant. Some of the rarer and more valuable minerals, as the agate, topaz, carnelian, and in unfrequent instances the diamond, have been found.

NEVADA.—Silver mining is the leading industrial pursuit of this state. The average monthly yield of the mines in the districts of Virginia, Gold Hill, Reese river, Esmeralda, and Humboldt, during the first nine months of 1865, was $1,331,555. Of this amount the greater part was extracted from a lode near Virginia City, in the western part of Nevada, where there is a ledge of ore running along the side of a mountain for three miles, with a width of fifty to one hundred feet, having a depth as yet unascertained. Over thirty companies have been working the same. The most prominent one of these has mined to the depth of eight hundred feet. Prior to April, 1866, the product of this lode was valued at $51,380,588; since then it is understood that fourteen millions more have been extracted. The bullion shipped from Virginia City and Gold Hill districts by express, during 1866, exceeded the shipment of the previous year by $2,074,174.

The mineralogist of the California state geological survey has expressed an opinion, supported by many scientific men, that the lode referred to is a fissure vein of extraordinary width and productiveness, and that ore will be found as deep as it is profitable to extend underground operations. The extension of railroad communications to such localities will render profitable the extraction of a low grade of ore with a fair margin of profit, adding $5,000,000 to the annual product of these mines.

Montana Territory.—The Surveyor-General of Montana Territory makes the following report to the General Land Office, in regard to mining in the district:

Coal.—Has been found on the Big Hole river, about sixty miles from Bannock City; in Jackass gulch, on the east side of the Madison river; and at Summit district, near Virginia City. These are all bituminous, and the seams do not exceed three or four feet in width, as far as known. Coal also exists on the head of the Yellowstone river. Brown coal, or lignite, is found in great quantities on the banks of the Missouri and Yellowstone rivers, valuable as common fuel, but of no great value for manufacturing purposes. It is also found on the head waters of the Teton and Marias rivers, branches of the Missouri.

Iron.—A deposit of iron ore has been discovered on Jackass creek, a tributary of the Madison river, but its extent is unknown. It is supposed to be valuable.

Copper.—There is a group of copper leads along the Muscleshell river, believed to be valuable. The lodes generally run east and west, and assayers have detected gold, in small quantities, in specimens examined. The width of vein is from three to four feet. Some recent discoveries of placer copper have been made on Beaver creek, near Jefferson City, which show some splendid specimens.

Silver Mines.—The first discovery of silver mines in the territory was made by Professor Eaton, of New York, on Rattlesnake creek, opposite the town of Argenta. The mineral was argentiferous galena. About the same time silver was discovered on the head of Prickly Pear creek, above Beavertown. Subsequently lodes containing both gold and silver were discovered at and near Virginia City, in the Madison range of mountains, on the Jefferson, Prickly Pear, Ten Mile, and Boulder creeks, and also in the vicinity of Helena The most recent discovery is on Flint creek, a branch of Hell Gate river. On the head of this stream a district has been found abounding in silver lodes, the assays of which have been of astonishing richness

Placer Mines.—The first placer mines worked in this territory were found on the Hell Gate river in 1862. In the fall the mines at Bannock were discovered. In May, 1863, the mines on Alder gulch, where Virginia City now stands, were discovered, and an immense impetus given to prospecting, and about $20,000,000 have been taken from there since that time. About a year afterward mines were discovered on Prickly Pear, where Helena now stands.

Valuable mines were discovered on the eastern side of the Missouri, and immense sums have been taken from Confederate gulch and Montana bar. Diggings were struck on the Big Blackfoot in 1865, which have produced largely.

All the gulches are on the head-waters of the Missouri, Columbia, and Yellowstone rivers, and are generally contained within the parallels of 45° and 47° 30′ and the meridians 110° and 114° west longitude. The number amounts to hundreds, and almost every day increases it.

Gulch mining is attended with many difficulties in this country. The season is short, and the gold generally found on the bed rock, often fifty or sixty feet from the surface. When provisions and labor become cheaper many gulches will be worked that at present are untouched. They are known to contain gold, but prices at present are too high to yield a profit on them.

Large amounts of money have been expended, this season, in ditches and preparations for gulch mining next year, and fully fifty per cent. more gold will be taken out than has been the present season. I estimate this year's work at $20,000,000. To conclude, not a tenth part of the territory has been prospected.

Gold-bearing Lodes.—The first gold-bearing lode of this territory was discovered at Bannock in 1862, and called the Dakota. The surface indications were extraordinarily good, and gave a stimulus to prospecting which resulted in the discovery of many valuable leads in that part of the country. The results from these leads proving satisfactory, several mills were erected, and are now at work upon rock taken from them. Some of the largest and finest lodes of this section have been discovered this season, and the hope is entertained that quartz mining may prove successful in this the pioneer mining camp of Montana.

Several thousand lodes have been discovered in Madison county, many of which are in the vicinity of Virginia City.

There are a large number of mills in this county, either in successful operation or in process of erection, and the results thus far have been generally satisfactory Much capital has been embarked in mining enterprises, and the work has been prosecuted with an energy that attests the confidence of the operators and gives assurance of success. At Summit City, eight miles above Virginia City, near the head of Alder gulch, gold quartz mining is extensively carried on. The lodes are numerous and rich; several mills are in operation and others being built. From this point there is a succession of auriferous lodes to the foot of Alder gulch, a distance of ten or twelve miles. Fine ledges are also found on the west side of the Madison range, in Ramshorn, California, and Beran's gulches, while Mill creek and Wisconsin gulches afford good prospects. Some of the gold-bearing lodes of this region contain large quantities of argentiferous galena.

There is a valuable quartz district between Hot Springs creek and Meadow creek, on the east side of the Madison range, and still another to the north of it, on Norwegian gulch. At the Sterling mining district, in this section, there are many valuable lodes, and five mills in operation.

There are also mining districts on the waters of the Jefferson river, known as the Silver Star, Highland, and Rochester, in which lodes have been found of immense value, some of which, in their present undeveloped state, have sold for large sums. One, the Green Campbell, was bought by a New York company for $80,000.

There are several districts on the Boulder, Prickly Pear, and Flint creeks and Deer Lodge river, which have shown fine indications and are being worked to a considerable extent.

The district of gold mining now receiving a considerable portion of public attention is that around Helena, a great many of the lodes being situated on the Oro Firio and Grizzly gulches, to the southwest of the city, stretching along to the northward toward Ten Mile, connecting with that district and Blue Cloud.

The Union Lode, No. 2, is situated near Grizzly gulch, in the Owyhee Park district. It is being worked in several places, and promises all that could be wished. Recent crushings of ore have yielded seventy-two dollars to the ton.

On Ten Mile creek, a stream that flows from a source near the summit of the Rocky mountains, in a northeasterly direction, there is a fine lot of lodes, some of which have assayed a large percentage of gold, while there is an intermingling of silver. Careful assays prove these lodes to contain from $25 to $300 per ton of ore, and by the "working test" made in St. Louis, $240 per ton has been obtained from rock taken from within seven feet of the surface. The veins are generally firm and solid within a few feet of the surface; the ledges from five to thirty feet high.

Blue Cloud, a new district, about ten miles from Helena, on Ten Mile, is opening out well. Machinery is being erected, and developments rapidly made.

In addition to the many mills, there are scattered over the different portions of the country, wherever there are any promising lodes, a large number of arastras. They are a rude mill, constructed for the purpose of working quartz, and generally driven by water-power. Most of them do well, and yield handsome wages to their owners. Some are erected for the purpose of developing mines, rather than going to the expense of bringing machinery on to the premises too early, deeming it best to prove the value of one good lode rather than own many with no knowledge of their intrinsic wealth. The owners of lodes are generally anxious to procure government patents for their claims, and already there have been several applications filed. Next season, I have no doubt but a large proportion of the owners of quartz will take advantage of the mineral law to get titles to their mines.

The leads in Montana are generally better defined than in any other mining country in the world, and the singular freaks sometimes taken by them in other regions are less frequent here. The simplicity of the ores is a theme of general remark, and although sulphurets are often found, they are taken as an indication of richness, and their appearance looked upon as a promise of ultimate success.

IDAHO TERRITORY.—The Surveyor-General says:

Gold is found on the head-waters of all the rivers. Rich placer mines have been profitably worked for years on the Clearwater and Salmon rivers. Extensive placer and quartz mines are found on the Boise river and its branches, embracing several districts. Many rich quartz lodes of gold and silver have been discovered and partially worked; their future development depending upon the reduced cost of transportation and other expenses, which thus far have retarded the growth and prosperity of the country.

The quartz and placer mines of Owyhee county, situated in the southwest part of the territory, have proved to be eminently rich so far as developed. Some of the ledges are being worked with valuable machinery, repaying the capital invested, though at an enormous outlay. The quantity and quality of the ore already abstracted are favorable indications of their future wealth.

Several thousands of gold and silver quartz claims have been taken up and recorded, more or less prospected, but the heavy expenses under which the miners of this territory have labored, has, in general, prevented their successful development. The near approach of the Pacific railroad to the southern borders of the territory will materially reduce the cost of working the mines, when the resources of the country will be more favorably brought into notice.

IRON.

In 1810 there was produced in this country 54,000 tons of iron; in the commercial collapse of 1820 the aggregate declined to 20,000 tons; in 1830 it had arisen to 165,000 tons, and in 1840 to 315,000 tons; in 1842, under the operation of the declining duties of the "compromise tariff," it had fallen to 230,000 tons;

under the combined influence of enhanced protective duties and high prices in England, caused by the sudden expansion of railway construction, it had arisen, according to the estimate of Hon. R. J. Walker, Secretary of the Treasury, to 765,000 tons; it rose to 800,000 tons in 1848, and fell to 650,000 tons in 1849, continuing to fall until 1852, when the entire product did not exceed 500,000 tons; in 1855 it had arisen to 1,000,000 tons, an aggregate which it nearly or quite maintained up to 1860. During our late difficulties the production of pig iron arose to 1,300,000 tons. Of manufactured iron in 1864 we produced 283,560 tons of railroad bars, with a capacity of increase to 700,000 tons per annum.

The azoic belt of Lake Superior is the great iron region of the globe. Though yet undeveloped, it furnishes in the single county of Marquette, in the upper peninsula of Michigan, one-eighth of the iron produced in the United States. The iron trade of this region has advanced from an export of 1445 tons in 1855, to 235,123 tons in 1864. The facilities for making charcoal there favor an extended production of fine malleable iron, while the inexhaustible supply of coal will supply fuel for the cheaper kinds of iron production. Iron mountain, in Missouri, rising 228 feet above a base of 500 acres, presents a cone of 230,000,000 tons. It is thought that every foot beneath its base level will yield an average of 3,000,000 tons; at a depth of 180 feet the artesian auger is still penetrating solid iron ore. Pilot Knob, the base of which, 581 feet beneath the summit, is an area of 360 square miles, is known to be solid ore to a depth of 440 feet below the surface. The upper section of 141 feet perpendicular thickness contains 14,000,000 tons of ore. Shepherd's mountain, one mile west of Pilot Knob, is a mass of very pure magnetic and specular ore, rising to the height of 660 feet.

COAL.

The known deposits of coal in the United States transcend in extent and richness those of all the residue of the world combined.

In 1845 our coal area was stated at 133,000 square miles. It is now known to be over 200,000 square miles, or eight times the known available coal area of all the rest of the globe.

Of the American coal-fields the Pennsylvania anthracite, though one of the smallest in area, is now the most copious in production, and the most available to the commercial and industrial interests of the nation.

Of our aggregate coal product of 22,000,000 tons in 1864, near 10,000,000 tons were mined in the anthracite region of Pennsylvania. They represent a commercial value of $60,000,000.

The production of the Alleghany coal-field, in 1864, was as follows:

	Tons.
Pennsylvania	5,870,712
Ohio	1,000,000
Maryland	657,996
West Virginia	500,000
Kentucky	250,000
Tennessee	500,000
Alabama	300.000
Total	9,078,708

The northern coal-field, embracing about 13,000 square miles, lies wholly within the lower peninsula of Michigan. But little has been done for its development, its annual product not much exceeding 100,000 tons.

The great central coal-field occupies an area of 50,000 square miles in Indiana, Illinois, and Kentucky. Its extreme length is 350 miles, with a breadth varying from 150 to 200 miles. The produce of the central coal-field in 1864 was as follews:

	Tons.
Illinois	1,000,000
Indiana	500.000
Western Kentucky	250,000
Total	1,750,000

The western coal-field in Missouri and Iowa is, properly, a continuation of the great central. It occupies an area of 45,000 square miles, of which 21,000 are in Missouri and 24,000 in Iowa. Its product in 1864 was about 500,000 tons. The coals of Arkansas and Nebraska are but the thin western edges of the great western coal-field, as yet but partially developed. All of these coal-fields are parts of the Appalachian coal, or Mississippi system, embracing a total area of 190,000 square miles, from which were mined in 1864 a total of 11,428,708 tons. The Mississippi basin, embracing an area of 1,500,000 square miles, is eminently available, in all its wondrous agricultural and mineral resources, to the demands of industry and commerce.

GOLD AND SILVER.

Of the quantities of the precious metals already taken from the mines of the United States, different estimates have been formed, some placing the product of California alone, since the commencement of 1848, at over one thousand millions of dollars. The special commissioner for the collection of statistics of gold and silver west of the Rocky mountains estimates the product of

California, from 1848 to the end of 1865, at nine hundred millions, and that of the neighboring states and territories, including the province of British Columbia, at $100,000,000, making an aggregate of $1,000,000,000. To reach this result the manifests at the custom-house at San Francisco have been taken, amounting to $740,832,623, to which was added the sum of $45,000,000 for gold and silver in use as currency on the Pacific, with an estimate of $115,000,000 for jewelry and plate manufactured in California, gold dust carried to the Atlantic states and foreign countries by miners returning home, without passing through the custom-house, and for dust buried or concealed by miners at remote points. It is safe to assume the total yield of Nevada, up to the end of 1867, at $100,000,000; that of Colorado at $30,000,000; of Oregon and Washington Territory, $25,000,000; Idaho and Montana, each, $25,000,000; and Arizona, New Mexico, and Utah, $5,000,000. If the product of California, up to the end of the same period, be assumed as equal to $900,000,000, the total product of the western mines up to the first of January, 1868, will amount to $1,110,000,000, or, in round numbers, $1,100,000,000, of which $1,000,000,000 may be set down for gold, and $100,000,000 for silver.

As to the annual product of the mines, opinions are likewise divided, some claiming eighty and others a hundred millions.

In 1865 and 1866 a revenue tax of six-tenths of one per cent. was collected on all gold and silver bullion in lumps, ingots, bars, or otherwise as assayed, which in 1866 amounted to $499,455, indicating a total value of bullion assayed, upon which a tax was paid, of $83,242,551 in paper currency value; equal in gold value to $56,000,000. A considerable quantity of bullion doubtless escaped taxation, but it is not probable the amount was greater than a fifth of the whole quantity subject to a revenue duty.

Considerable quantities of these metals passed into manufactures without being previously assayed, and left the country in the form of dust by miners returning to foreign parts, or was shipped in the form of ore; and $5,000,000 may be set downas a contingent under these heads, making a total of $75,000,000, gold value, for the year 1866, of which $18,000,000 represent the silver product.

The amount deposited at the mints for the year 1866 was less than $32,000,000, gold value, the mint returns exhibiting about four-sevenths of the amount of assayed bullion produced during that year upon which a revenue tax was assessed and paid.

HOMESTEAD RULINGS.

A CASE was before the Commissioner, upon an application to cancel homestead entry, in which the testimony showed it to have been made while the claimant was serving in the army of the United States; that he was mustered out of the service in May, 1866; soon thereafter built a small shanty and commenced work upon the homestead, continuing to occupy the same, with frequent absences, though not at distant intervals.

It was held that the claimant had not abandoned his tract; yet the character of his improvement and manner of residence were not such as the law contemplates. The cancellation of this entry was declined on the evidence adduced, while a more complete compliance was called for under the law. It was insisted that the settler must put upon the land such a house as may answer for permanent residence—not merely a place of temporary resort in order to show his intention to comply with the law—and make the land what the statute intends—his actual homestead. A period of sixty days from the date of notice was allowed within which to complete his house and move therein, it being required at the expiration of that time that he should appear before the register and receiver and show by affidavit, supported by corroborative testimony, compliance with such requirements.

An entry has been presented in which the homestead was made and commuted before the expiration of five years from its date, and the point submitted was whether the settler could make another like entry. The sixth section of the homestead act of 1862 is specific in its declaration "that no individual shall be permitted to acquire title to more than one quarter section under the provisions of this act." Therefore it is held that when a party acquires title under any of the provisions of this act, his privilege is thereby exhausted.

Inquiry has been made whether persons employed in the military or naval service of the government may take homesteads under the amendatory act of 21st of March, 1864, and obtain title to the land, supposing their time of service to absorb all of the five-year period for settlement and cultivation.

The ruling is that actual settlement and cultivation of the land are required by law; hence title cannot be acquired unless the party, immediately upon discharge from service, enters upon the land, makes it his home, and cultivates the same as required by the original act of 20th May, 1862; actual settlement upon and cultivation being required in all cases.

In the second section of the law of 20th May, 1862, it is stipulated in regard to settlers that in the case of the death of both father and mother, leaving an infant child or children under twenty-one years of age, the right and fee shall inure to the benefit of the infant child or children; and that the executor, administrator, or guardian may sell the premises for the benefit of the infant heirs at any time within two years after death of the surviving parent, and in accordance with the law of the domicile.

The question has been made as to whether it is imperative the land shall thus be sold under the statute for the benefit of the heirs, or whether they can retain title under the original settlement.

The Commissioner rules that there is no objection under the general provisions of the law to the maintenance of settlement and cultivation on the part of the heirs, and the issue of the patent in their names at the expiration of the required time; yet in such case the minor heirs must continue to reside upon the homestead, and the settlement and cultivation of the same must be continued for their benefit.

In the case of a homestead settler who died unmarried, and whose father applied for preference right to take by ordinary purchase at $1.25 per acre, it is held that although privilege could not be granted, yet on satisfactory showing of the death of the settler, with proof of his previous settlement and cultivation, the land could be paid for at $1.25 per acre by the heirs of the deceased settler, under the eighth section of the act of 20th May, 1862, and entry made under that section in favor of the heirs of the decedent, following in this respect the rule prescribed by the second section of the act of 3d March, 1843, in regard to deceased pre-emptors.

HOMESTEAD COMMUTATIONS WITHIN RAILROAD LIMITS.

In the Secretary's decision of 22d June, 1866, as communicated in our circular of 25th August, 1866, the question discussed was the rights of pre-emption settlers to commute their filings on odd sections after the railroad withdrawal had been made. The Secretary then decided "the homestead settler's right attaches only from the date of entry," the pre-emptor from the date of his actual personal settlement; hence, if the commutation is not made prior to withdrawal, it cannot be made afterward, the railroad grant taking effect immediately upon the abandonment of the pre-emption for the homestead.

By parity of reasoning this principle will apply to the *even sections* within the granted limits, in the matter of price; hence. if a party fail to commute prior to the increase from minimum to double minimum, he cannot commute afterward at the rate of $1.25 per acre, because upon the relinquishment of the pre-emption claim, the double minimum immediately attaches, and he will be required either to prove up as an original pre-emptor at $1.25 per acre, or in commuting to restrict the homestead claim to 80 acres.

Inquiries have been presented as to what is required of heirs at law desirous of making payment under the 8th section of the act.

The requirements are the production of evidence of heirship, with proof that the deceased settler had fully met the requirements of the statute by residing upon and cultivating the tract up to the date of decease; and, further, that the improvements had been continued and residence maintained by the heirs upon the homestead after the death of the settler; or, in case residence and cultivation had not been continued, proof that at the date of the application a sufficient time had not elapsed since the decease to work a forfeiture of the claim.

Instances have occurred in which a widow has made a homestead settlement and thereafter marries a person who likewise made a similar settlement on another tract. It is ruled that the parties may elect which tract they will retain for permanent residence, and that on proving up settlement under the 8th section of the act of May 20th, 1862, the title to the remaining entry may be perfected by the original settler.

THE RIGHTS OF FOREIGNERS IN RELATION TO THE ACQUISITION OF TITLE TO PUBLIC LANDS.

As aliens cannot acquire valid titles to real estate under the pre-emption and homestead laws, the privileges of which are restricted to citizens, or those who have declared their intention to become such, it is important that foreigners seeking identification with the American community, should be advised of the legal steps necessary to acquire citizenship. To that end it is submitted that any free white alien, over the age of twenty-one years, may at any time after arrival declare before any court of record having common law jurisdiction (with a clerk or prothonotary and seal) his intention to become a citizen, and to renounce forever all foreign allegiance. The declaration must be made at least two years before application for citizenship (U. S. Statutes, vol. ii. page 153, and vol. iv page 69.)

At the expiration of two years after the declaration, and at any time after five years' residence, the party desiring naturalization, if then not a citzen, denizen, or subject of any country at war with the United States, should appear in a court of record, and there be sworn to support the Constitution of the United States and renounce foreign allegiance. If he possessed any hereditary title or order of nobility, such also must be renounced, and satisfactory proof produced to the court by the testimony of witnesses, citizens of the United States, of the five-year residence in the country, one year of which must be within the state or territory where the court is held, and that during the five-year period he was a man of good moral character and attached to the principles of the Constitution; whereupon he will be admitted to citizenship, and thereby his children, under twenty-one years of age, if dwelling in the United States, will also be regarded as citizens. (U. S. Statutes, vol. ii. page 155.)

Where the alien has made his declaration and dies before being actually naturalized, the widow and children become citizens of the United States and entitled to all rights and privileges as such, upon taking the prescribed oaths. (U. S. Statutes, vol. ii. page 292.)

Any free white alien, being a minor, and under the age of twenty-one years at the time of arrival, who has resided in the country three years next preceding his majority of twenty-one years, may, after reaching such period and on five years' residence, including the three years of his minority, be admitted to citizenship without a preliminary declaration of intentions, provided he then makes the same, averring also on oath and proving to the court that for the past three years it had been his intention to become a citizen; also showing the fact of his residence and good character. (U. S. Statutes, vol. iv. page 69.)

Children of citizens of the United States born out of the country are deemed citizens, the right not descending, however, to persons whose fathers never resided in the country; and any woman who might legally be naturalized, married, or who shall be married to a citizen of the United States, is held to possess citizenship. (U. S. Statutes, vol. x. page 604.)

An alien twenty-one years of age and over, who has enlisted, or shall enlist, in the regular or volunteer armies of the United States, and be honorably discharged, may be admitted to citizenship upon his simple petition and satisfactory proof of one year's residence prior to his application, accompanying the same with proof of good moral character and honorable discharge. (U. S. Statutes, vol. xii. page 597.)

FOREIGN TITLES WHICH ORIGINATED UNDER THE FRENCH, SPANISH, BRITISH AND MEXICAN GOVERNMENTS.

In acquiring territory the United States have stipulated in different treaties for the recognition and protection of private property. This has been done not only as a measure of justice, but in coincidence with the public law.

CALIFORNIA.—Under the several acts of Congress for the settlement of Spanish and Mexican claims in that state, surveys have been reported in three hundred and sixty-seven cases, covering five million six hundred and ninety thousand five hundred acres; and of these patents have been issued for two hundred and seventy-five claims, embracing four million three hundred and sixty-three thousand three hundred acres.

FLORIDA, LOUISIANA, AND MISSOURI.—The act of Congress approved March 2d, 1867, continues in force for three years the provisions of the statute of June 22d, 1860, "for the final adjustment of private land claims in the states of Florida, Louisiana, and Missouri, and for other purposes." That act constituted the registers and receivers of the several land offices in Florida, Louisiana, and the recorder of land titles at St. Louis for the state of Missouri, commissioners to hear and decide, under instructions from the General Land Office, all matters respecting claims to land within their several districts. The law confers power upon them to receive only such claims as are founded on *written* grants, and hence interdicts action upon any interest founded merely on ancient settlement, when the same is unaccompanied by paper title from the authorities of the former government.

These statutes authorize the reception and action upon such claims for tracts within the several districts as have emanated from any foreign government, bearing date prior to the cession to the United States of the territory out of which the states were formed, or during the period when any such government claimed sovereignty or had the actual possession of the district or territory in which the lands so claimed are situated. This warrants them in receiving and acting not only upon claims which originated under the former governments while the authorities exercised the granting power *de jure*, before the cession of the country, but also allowed claims to be received which were made by the Spanish authorities while they were in actual occupancy of territory as the government *de facto*. Thus, for example, Spain parted with authority over the province of Louisiana by the secret treaty of 1800 at San Ildefonso, when that power ceded Louisiana to France. During the period that elapsed from that time to the cession to the United States in 1803, by Napoleon, the Spanish authorities exercised the granting power; and so, several years subsequent to 1803, Spain, while in occupancy of the ancient province of Louisiana between the Iberville or Manchac and the Perdido, continued to make land concessions; and during this period the grants were, of course, those of the government *de facto*. Titles of this class stood excluded by the ruling of the Supreme Court of the United States in the case of Foster and Elam *vs.* Neilson (2 Peters's Supreme Court of the United States), in which an elaborate decision was rendered by the Chief Justice against their validity under the then existing laws and treaties. Now, by the force and effects of the said acts of 1860 and 1867, a status is given to claims founded on titles from *de facto* governments after the authority *de jure* had passed from them, a principle being thus legislatively recognized which had not previously been done nor admitted in the judicial rulings of the Supreme Court of the United States.

Numerous claims that had their origin from governments which preceded the United States in sovereignty on this continent, have been finally confirmed by boards of commissioners, and by judicial decrees; but the greater number have been affirmed by direct legislative acts operating upon official reports submitted from time to time to Congress, from an early period.

In multitudes of cases parties in interest seem to have rested, as sufficient evidence of their right, upon the decrees or acts of confirmation, and actual possession, and hence the apathy in that respect which has existed in not applying for patents or complete titles as authorized by acts of Congress.

The General Land Office, however, is prepared to meet all lawful demands in the way of conferring such complete titles, where the terms of confirmation shall have been fully met by the rendition of authenticated plats of surveys, accompanied by the patent certificates, the statutory provisions generally making such instruments the necessary basis of patents from the United States.

LEGISLATION IN RELATION TO DISCONTINUANCE OF SURVEYING DISTRICTS.

An act for the discontinuance of the office of Surveyor-General in the several districts, so soon as the surveys therein shall be completed, was approved June 12th, 1840. (Section 1, Statutes, vol. v. p. 384.)

Another law was passed January 22d, 1853 (Statutes, vol. 10, p. 152, amendatory of that of June 12th, 1840), providing that the field-notes, maps, records, and papers, may be turned over to the state authorities when a Surveyor-General's office is discontinued; the amendatory statute clothes the Commissioner of the General Land Office with authority to act *ex-officio* as Surveyor-General, the statute stipulating in behalf of the United States for free access to the archives after the same shall have been delivered to the state.

Pursuant to those enactments the archives were delivered, years ago, to the authorities of Ohio, Indiana, Michigan, Alabama, Mississippi, and more recently to Missouri and Wisconsin, the public surveys having been completed in those states. The records for Arkansas and Illinois, where the field-work has been finished, are awaiting, as preliminary to delivery, the state legislative acceptance, according to the terms presented by acts of Congress. In Arkansas the completed archives, in 1859, were placed for safe-keeping with the register at Little Rock; other records connected with the surveys in that state are in charge of the General Land Office, while the surveying records for Illinois remain in the custody of the recorder of land titles at St. Louis. In Iowa, where the field-work is completed, the surveying records are at Dubuque, in anticipation of the requisite state legislative acceptance of the same.

The archives for Louisiana are in possession of the United States land office at New Orleans, and those for Florida are at Tallahassee.

LEGISLATION RESPECTING THE DISCONTINUANCE OF LAND OFFICES.

The law of June 12th, 1840 (second section Statutes, vol. v. p. 384), orders that whenever the quantity of land remaining unsold in any district shall be less than one hundred thousand acres, the district shall be discontinued, and the land unsold made subject to sale at the land office most convenient to the place in which the land office shall have been discontinued.

The 7th section of the act of September 4th, 1841 (Statutes, vol. v. p. 455), declares that land districts may be continued, if the quantity of land unsold does not equal one hundred thousand acres, should such continuance be required for public convenience, or in order to close the land system in a state.

The law of March 3d, 1853 (Statutes, vol. x. p. 194), provides that land offices may be annexed to adjacent districts by the President, whenever the cost of collecting revenue from sales of public lands in a district amounts to one-third of the whole revenue there received, if, in his opinion, the consolidation is not incompatible with the public interest.

By the act of March 3d, 1853 (Statutes, vol. x. p. 244), authority is conferred upon the chief executive to change the seats of land offices.

The first section of the act of February 18th, 1861, vol. xii. p. 131, authorized the register and receiver of the consolidated office, at Boonville, Missouri, to receive fees for certain services. The second section of that law allows office rent and clerk hire, if sanctioned by the Secretary of the Interior. The third section extends the provisions of that statute to all consolidated offices.

In the fifth section of the law of 30th May, 1862, vol. xii. p. 409, it is declared that upon recommendation of the Commissioner of the General Land Office, approved by the Secretary of the Interior, the President may direct the discontinuance of any district, and the transfer of its business and archives to any other land office within the same state or territory.

By the seventh section of the act of 26th July, 1866, Laws, p. 252, the President is authorized, in reference to mining interests, to establish additional land districts, and appoint officers under existing laws, whenever he may deem it necessary for the public convenience. In executing the provisions of the law, registers and receivers, under this act, are held by the General Land Office as possessing powers coextensive with all other interests connected with the disposal not only of mineral but agricultural lands.

Under the above-mentioned act, additional land offices have been created and established as follows:

At Sacramento, California, district composed of the counties of Sierra Nevada, Placer, El Dorado, Amador, Calaveras, Alpine, and Sacramento.

At Austin, Nevada, district composed of the county of Lander.

At Belmont, Nevada, district composed of the counties of Nye, Esmeralda, and the unorganized county of Lincoln.

At Lewistown, Territory of Idaho, district embracing all that part of the territory lying north of the Salmon river range of mountains.

At Fair Play, Territory of Colorado, district composed of the counties of Lake, Park, and Summit.

Under the act of 27th of June, 1866, Laws, p. 77, a land office has been established at Boise City, Idaho Territory, the district embracing all that part of the territory lying south of the Salmon river chain of mountains.

Under authority of the act of March 2d, 1867, Laws, p. 542, a land office for the Territory of Montana has been established at Helena, and the site of a land office for the Territory of Arizona has been fixed at Prescott.

In accordance with the provisions of the act of 3d March,

1853, the President, under date of 16th February, 1867, directed that the office for the sale of public lands, in the southwestern district of Alabama, be removed from St. Stephen's to the city of Mobile, in that state.

In April, 1867, an order was given for the discontinuance of Elba land district, in the state of Alabama, and the vacant public lands therein were made subject to sale at Montgomery.

By notice, dated 8th May ultimo, the land offices at St. Augustine, Newnansville, and Tampa, in the state of Florida, were discontinued, and the vacant lands in the several districts were made subject to sale at Tallahassee, thus consolidating all the land offices in the state at the capital.

INDEX.

www.ingramcontent.com/pod-product-compliance
Lightning Source LLC
LaVergne TN
LVHW050526100826
845148LV00002B/456